The Impact of Identity in K–12 Mathematics:

Rethinking Equity-Based Practices

Expanded Edition

Julia Maria Aguirre
University of Washington–Tacoma

Karen Mayfield-Ingram
University of California, Berkeley

Danny Bernard Martin
University of Illinois Chicago

NCTM

NATIONAL COUNCIL OF
TEACHERS OF MATHEMATICS

more**4**u
www.nctm.org/more4u
Access code: IIM15951

A Publication of

NCTM ® | **NATIONAL COUNCIL OF**
TEACHERS OF MATHEMATICS

1906 ASSOCIATION DRIVE | RESTON, VA 20191-1502

Cover Design: Kirsten Ankers
Layout: KnowledgeWorks Global Ltd.

*****

The National Council of Teachers of Mathematics advocates for high-quality mathematics teaching and learning for each and every student.

****

Recommended citation:
Aguirre, J., Mayfield-Ingram, K., and Martin, D. (2024). *Impact of Identity in K–12 Mathematics Rethinking Equity-Based Practices Expanded Edition*. Reston, VA: National Council of Teachers of Mathematics

ISBN: 978-1-68054-048-2
ISBN (ebook): 978-1-68054-049-9
Library of Congress Control Number: 2023951842

****

Printed in the United States of America

Contents

Preface

Over a decade ago, we offered our book, *Impact of Identity in K–8 Mathematics: Rethinking Equity-Based Practices*, to invite critical reflection on mathematics teaching and learning that influences students' mathematical identities. This expanded edition represents an extension of the ideas in the first edition to the high school context. Thus, we offer a new focus on the *Impact of Identity in K–12 Mathematics* to encourage critical dialogue and reflection about making mathematics education more meaningful, liberatory, and just. As it was then and is now, our goal is not to generate discourse *about* teachers. Instead, it is to engage in conversations *with* teachers, raising issues that we have seen in our individual and collective classroom experiences as teachers and researchers and that many teachers will recognize from their own experiences.

We have chosen to focus on teacher reflection and practice in the context of mathematics learning and identity development. That is, we have chosen to focus on understanding how teachers help students become empowered mathematics learners and how students come to see themselves in relation to mathematics learning, both inside and outside the classroom. Teaching is complex professional work that requires ongoing reflection on curriculum and content, as well as self-reflection—reflection about children and families, reflection about the role of mathematics in the lives of children and families, and even reflection about routine, everyday practice. Our commitment to teacher reflection is grounded in our desire to help teachers understand, appreciate, and draw on the backgrounds of students as strengths to further their mathematical development—especially those students who have not had equitable access or opportunity to learn mathematics.

In the first edition of this book, we noted the increasingly political environment of education reform and the impact on mathematics teaching and learning. Over the last ten years, this has not waned. The work of teachers is still subject to scrutiny, and we continue to hear laments about the dismal state of U.S. mathematics education. The move toward the Common Core State Standards has not resulted in significant advances in achievement or equity. COVID-19 has further exposed inequities in our educational system and has brought to light the difficult working conditions of many teachers. During the pandemic, teachers, like other essential workers, put politics aside and courageously continued to do their jobs teaching children while trying to protect themselves and their families. Unfortunately, politics did not press pause, making working conditions for teachers challenging. Many teachers have left the profession. In this new edition, we examine the possibility that the politics of reform, often rooted in incrementalism and linked to larger political projects, lean more toward preserving the status quo instead of resulting in the kind of transformative change that prioritizes liberation and justice for those who continue to be oppressed. We call on teachers to question these reforms and ask themselves whether reforms support or diminish the work of teachers and the humanity of their students.

Significantly, in this expanded edition, we build on these realizations about reform to extend our focus on equity. We realize that working toward equity in mathematics education and elsewhere is necessary but not sufficient to move beyond the status quo. In the last ten years, social events across the globe, but especially in the United States (e.g., police killings of unarmed Black people; increasing anti-immigrant rhetoric and violence; forced separations of families seeking asylum; missing and murders of Indigenous women; anti-Asian hate crimes; #MeToo movement; increases in domestic violence; rollbacks in civil rights by overturning Supreme Court decisions protecting voting rights and people's bodies; banning books related to race, gender identity, and sexuality; and political insurrections with an attack on the U.S. Capitol building), remind us that all our equity-focused efforts take place within the oppressive systems and political projects of white supremacy, antiblackness, and xenophobia. Equity unfolds in the system as it is and not the systems we desire. Equity alone does not result in liberation, freedom, or justice. Within these systems and projects, the positive mathematics identities we stress in this book are just one tool to combat these forces. We continue to call on teachers to examine their beliefs and practices and to work with parents, caregivers, and communities to support our children as whole human beings.

In the expanded edition of this book, we acknowledge the dual role that mathematics has played as a gatekeeper and gateway to various opportunities in society, and we recognize that mathematics has often been used to make judgments about intelligence. We call on teachers to reflect on these uses of mathematics and develop an asset-based approach that recognizes the ordinary brilliance, creativity, and curiosity children bring to learning spaces. Further, because students do not come to school without the influences of their homes, communities, and cultures, we ask teachers to reflect on how the multiple identities that students are developing can influence their mathematics learning. Yet, we do not confine the conversation on mathematics learning to students. We ask teachers to reflect on their own learning experiences and how those experiences have shaped their senses of themselves as doers of mathematics, how their views of themselves as mathematics learners affect their classroom practice, and how they navigate the unjust education systems in which they work.

This expanded edition consists of twelve chapters divided into three parts. It is designed to help teachers move beyond an awareness of the need to reflect on their teaching to a commitment to take action and transform their teaching through equity-based practices. An epilogue offers final reflections that go beyond equity toward a more humanizing and liberatory mathematics education. A list of discussion questions appears at the end of each chapter to promote further dialogue and self-reflection. Readers can also use the access code on the title page to find specific reflection tools and activities at the More4U website (www.nctm.org/more4u) from the National Council of Teachers of Mathematics (NCTM).

Part 1 focuses on mathematics learning and identity. It is expanded to include two new chapters designed to cultivate critical reflection about the political projects and oppressive structures that adversely affect mathematics identity. Chapter 1 expands upon important themes related to the purposes of learning mathematics, equity, and shifts needed to attend to learning and identity. Chapter 2 discusses the connection between mathematics learning and student identity. This chapter discusses why teachers must attend to this linkage to empower young mathematical minds. Chapter 3 shifts that identity focus to teachers, calling for reflection on the impact of one's own mathematics learning identity on instructional beliefs and classroom practice. Chapter 4 critically examines the political contexts of standards and standard-based practices, asking us to consider the fundamental question: Are these standards and standards-based practices worthy of my students? Chapter 5 explores ways to disrupt deficit-based thinking and structures that prevent joyful mathematics learning.

Part 2 describes five equity-based instructional practices designed to strengthen mathematics learning and positive mathematics identity:

- Going deep with mathematics
- Leveraging multiple mathematical competencies
- Affirming mathematics learners' identities
- Challenging spaces of marginality
- Drawing on multiple resources of knowledge

The chapters in this part use classroom vignettes to bring these practices to life. In this expanded edition, we include instructional examples and student stories situated in high school mathematics. We also include a teaching analysis tool that can help facilitate individual reflection and collaborative feedback. Chapter 6 describes a middle school mathematics teacher's practice that cultivates mathematical agency and empowers students to examine racial profiling claims by using mathematical evidence. Chapter 7 focuses on ways that an elementary teacher builds on students' strengths to foster their engagement with mathematics and thus support their learning of mathematics. Chapter 8 focuses on seeing brilliance and creating access for Black girls in high school mathematics. Chapter 9 shines the spotlight on assessment. Drawing on the five equity practices, this chapter specifically discusses the role of meaningful feedback on typical classroom "tests"—feedback that can deepen students' learning and help develop their sense of themselves as mathematics learners. This chapter also describes equity-focused grading policies that affirm students' mathematical identities. In addition, the chapter highlights ways that teachers can learn to recognize various experiences and knowledge that students bring to bear on assessments and that demonstrate what they know and can do.

Part 3 focuses on the importance of engaging families, caregivers, and communities as true partners in supporting mathematics learning and positive mathematics identity development in elementary and secondary settings. We acknowledge that families are multifaceted and unique with various people in parental and caregiving roles including extended family members, foster families, grandparents, cousins, older siblings, and so on. We recognize and affirm that many cultures embrace a community-oriented and collaborative approach to caregiving with more expansive and inclusive family structures to support the well-being of children. Chapter 10 discusses routine strategies such as family newsletters and conferences, which can be enhanced to strengthen relationships with parents and caregivers and effectively communicate the teacher's mathematics vision and the student's progress. Chapter 11 moves beyond the classroom walls to highlight ways in which teachers and schools can partner with parents, caregivers, and communities to support mathematics learning and provide complementary resources to help children learn mathematics. Chapter 12 focuses on communicating with families about mathematics education in the middle school and high school years.

The epilogue offers final reflections about important ideas in the book and ways for teachers to transcend equity-based practices for liberatory and humanizing purposes that strengthen mathematics learning and positive mathematics identity development for children.

In writing this book, we have drawn on our own experiences in K–12 classrooms; our experiences as and with caregivers, teachers, and children; and our experiences as scholars and teacher educators in preservice and in-service contexts. Each of us has a lifelong history of commitment to issues of equity and justice in mathematics education, focusing on uplifting historically marginalized students and families. Our professional experiences and practices have been devoted to empowering students with mathematics to help them realize a full range of educational and life opportunities.

Our own journeys in mathematics, including being identified by teachers for accelerated tracks in mathematics, participating in academic enrichment programs for people of color and women, experiencing undergraduate and graduate study in mathematics and education, and being mentored in the areas of mathematics and mathematics education, have also shaped the writing of this book. We have known firsthand the impact of teachers on our lives in relation to mathematics learning.

Moreover, writing this book has been shaped by our own racial and cultural identities as well as the ways that we have come to see ourselves as learners and doers of mathematics because of those identities. Racially, ethnically, and culturally, for example, we self-identify with particular social categories (African American, Latina, multilingual, multiethnic, biracial) that have been marginalized in societal and school settings, especially with respect to mathematics. These social categories are both personal and political and have evolving histories and meanings, and we acknowledge the sociopolitical and power implications of these identifications.

We acknowledge that these categories and others—for instance, "Black," "white," "Latinx," "Asian," and "Native American"—are negotiated, and each of us makes our own sense of what these terms mean and whether we choose to make them our own. We also acknowledge the shared history and experiences of group members despite different labels. For example, we use the labels "African American" and "Black" to encompass the diverse ways that group members of the African diaspora self-identify and are identified by others. We use "Latinx" rather than "Latina/o" as a gender-inclusive pan-ethnic label to express the cultural and political solidarities of people who are descendants of or natives of a Western hemisphere country south of the United States, including Mexico and the countries of Central and South America and the Caribbean (Hurtado and Gurin 2004). "Latinx" acknowledges Indigenous, African, and European ancestries that the label "Hispanic" does not. The categories "white" and "Asian" also include multiple identifications. In this book, we ask educators to recognize and critically reflect on how all these terms are used in school discourses related to mathematics.

We hope that this book, in which we bring to bear our professional and personal experiences, will be an essential resource for teachers, teacher educators, and education researchers interested in teacher development, equity, learning, and identity. We trust that it will also prove useful to parents and caregivers, as well as school administrators and instructional coaches who wish to support teachers in the teaching of mathematics. Most importantly, we hope that we have succeeded in shaping this book in such a way that it will push our collective thinking and practice to give our nation's youth a better preparation for learning mathematics and developing positive mathematics identities that will advance their own educational, career, and life opportunities. We hope that the examples presented in this book will resonate with teachers and provide opportunities for them to reflect critically on their beliefs, practices, and systems that impact their essential work.

Acknowledgments

We are grateful for and feel very fortunate to have been given the opportunity to write this expanded edition of this book. We owe a great deal of thanks to NCTM's

Publications Committee for supporting this expanded edition. We appreciate the opportunity to include new chapters that focus on math identity development at the high school level. We are especially grateful to Myrna Jacobs, Christine Noddin, Stephanie Levy, Scott Rodgerson, and Mary Donovan, NCTM publications team, for their patience and support for the first and expanded edition of this book. We offer special thanks to our teacher leader colleagues Marcie Baril, Andy Coons, Kristen Faust, Tracy Fischer, Desiree LeSage, Shelley Rafter, Mary Richards, Susan Sabol, Alexi Salvador, and Niral Shah for their insights, examples, and constructive critiques that helped us to think through some of the issues raised in this book. We also thank our families for their ongoing support and inspiration.

We dedicate this book to all young students, past, present, and future, who continually teach us what is necessary to help them learn and grow mathematically. And to the educators, families, caregivers, and communities who, in the face of many challenges, demonstrate their dedication, love, and support for young people.

Julia's special acknowledgments

I thank Karen Mayfield Ingram and Danny Martin for our ten-plus years of collaboration and laughter. Danny, you have been an inspiration and guiding light ever since graduate school and the functions group. Thank you for your generosity and critical insights. Karen, thank you for the opportunity to learn from and with you over these years. I've appreciated the space to dialogue, ask questions, and process emotions when navigating life. I also thank mí amiga, Rochelle Gutiérrez. We have known each other for a long time. Thank you for believing in me and inspiring many ideas for this book. To my parents, Ricardo and Marlinda Aguirre, thank you for your love and faith in me. I am grateful to my children, Alesandra and Joaquín, who are now young adults navigating the world. Thank you for teaching me how to be a better person, parent, and educator. I love you forever. To my wonderful husband, Jade, thank you for your love, delicious meals, and unwavering support through the years. I appreciate you loving all of me.

Karen's special acknowledgments

I am grateful to the women and men in my family for modeling what it means to believe in and support the brilliance of Black children. I appreciate and love the brilliance of my daughter, Janelle, for her valuable feedback and perspective. To my loving husband, LaRon, who supports my crazy hours and the unbalanced juggling of family and work with calm and ease, thank you, sweetheart. To my colleagues, Julia and Danny, I am grateful for your unwavering support and friendship throughout our book journey. Your commitment and boldness in advocating for our children and families gives me strength. I have learned so much from each of you and am so thankful for the opportunity I've had to work and get to know you both.

Danny's special acknowledgments

A special shout out to my eleven-year-old son, Ishaan, and my eighty-year-old mother and eighty-four-year-old father. Their lives serve as reminders of the past and the future and keep me motivated to do this work.

Part 1

Rethinking Mathematics Learning, Identity, and Equity

Part 1 provides foundational concepts to understand the individual, interactional, and social forces that impact mathematics identity development. We believe that focusing on the relationships between mathematics learning and identity can help teachers reflect on their beliefs about students, mathematics learning, equity, and justice. Examples are provided from elementary and secondary contexts. In Chapter 1, we ask teachers to reflect on the purposes of learning mathematics and the kinds of experiences that can lead students to develop positive and negative math identities. To spark discussion, we reiterate our definition of equity offered in the first edition. We then expand that definition toward a more humanizing and liberatory mathematics education.

Chapter 2 examines how a rich and complex view of students includes attention to the multiple identities they are in the process of developing on the basis of their experiences in school and outside of school. We give particular attention to how these experiences shape and are shaped by the students' mathematics identities and how teachers can leverage students' multiple identities to promote progress in mathematics, especially for historically marginalized students. Teachers are encouraged to reframe students' identities in ways that move beyond stereotypes and deficit views. We stress that students are active participants in their mathematics learning and should be encouraged to engage in multiple forms of mathematical agency.

Chapter 3 explores a variety of experiences that contribute to the development of what we call a "mathematics teacher identity." We present six case studies of new teachers, all of whom discuss their own experiences as learners of mathematics and aspects of their identities that were relevant in those experiences. Professional experiences and demands also shape mathematics teacher identities. Decisions and mandates about what content gets taught, to which students, and by which teachers, shape mathematics identities for teachers and students. We give particular attention to mandates calling for "algebra for all" and how those mandates can challenge the identities of elementary and secondary teachers.

Chapter 4 critically examines the roles of standards and standards-based practices on math identity development and mathematics education. We reframe standards and standards-based practices as sources of mathematical, social, and knowledge empowerment for students and their teachers. This reframing centers children's humanity and brilliance by asking a fundamental question: Are these standards and standards-based practices *worthy* of our students? This focus shifts the discourse toward a more humanizing and liberatory mathematics education.

Chapter 5 introduces a professional learning activity designed to help disrupt and reframe pervasive deficit-based beliefs and thinking, as well as institutional practices that promote hierarchies among students, such as curricular tracking. We discuss specific strategies teachers can use to dismantle such beliefs and practices.

We encourage teachers, individually and in groups, to engage with the discussion questions at the end of each chapter. These questions provide opportunities to reflect on their own identities as well as the identities of their students. In the following chapters, all names of students, teachers, and schools in the vignettes and examples are pseudonyms unless otherwise noted.

Chapter 1

What Mathematics? For Whom? For What Purposes?

Throughout this book, we encourage all teachers to reflect on the three questions in this chapter's title: What mathematics? For whom? For what purposes? We raise these questions because they strike at the core of equity concerns in mathematics education and remind us that school mathematics simultaneously serves as a gateway and gatekeeper for various opportunities in and out of school. As expressed in policy and reform documents, these opportunities include access to advanced courses, entrance to college, and access to math-dependent college majors and careers. Teachers play a crucial role in deciding which students will or will not have access to these opportunities (e.g., Faulkner et al. 2014).

Mathematics also provides a critical lens for discerning patterns and making sense of quantitative information (e.g., data on COVID-19, data on police shootings, political polling information) that we experience in the world every day. However, some students are never given the opportunity to engage with mathematics in rich and meaningful ways that emphasize critical thinking and problem solving. Moreover, educators often use mathematics assessments to make definitive judgments about students' competencies and abilities. Such judgments can follow students throughout their academic careers and have a long-lasting impact on how they see themselves as doers of mathematics (Boaler 2002, 2008; Davis and Martin 2018; Gholson and Wilkes 2017; Jackson 2009; Langer-Osuna and Esmonde 2017; Larnell 2016; Martin 2000, 2009b; Miller-Cotto and Lewis 2020; Spielhagen 2011; Wood 2013).

We contend that deep, meaningful reflection on these questions will require teachers to examine their beliefs about learners, learning, and mathematics content, as well as their everyday teaching and classroom practices. This work also demands that teachers closely examine their own mathematics learning experiences and how these shape their instructional visions and classroom practices. Furthermore, by thinking about how best to support students, teachers may need to reexamine their beliefs and retool their practices, not only to engage students more effectively in learning mathematics but also to partner more successfully with families and communities to support learning in and out of school.

This book provides examples, concepts, and reflection tools that teachers can use to build richer perspectives on issues of equity and justice within the context of their routine everyday classroom practices. Key to developing these richer perspectives and practices is attending to issues of identity with and through mathematics—that is, understanding who students are, who they are becoming, and who they want to become. It also means understanding how students are positioned as mathematics learners by what others—teachers, peers, parents, or caregivers—say about who they are, such as who gets identified as "good at math" and who does not.

Consider the following vignette involving a young student, Baye, his father, and his mathematics teacher. The vignette highlights several issues related to student beliefs about mathematics, teacher practice and influence on those beliefs, and parental support for mathematics learning.

Baye is a third-generation Korean American sixth grader in his first year at Crestmont Middle School, located in a midsized city in the western United States. He is a rising star on the local track team and volunteers as a faith mentor for younger children at his local church. Although he has many leadership qualities and a strong preference for science, he struggles with learning mathematics.

Ms. Carlson is Baye's mathematics teacher. She believes strongly that all students can be successful in mathematics if given the right content and exposed to the right pedagogy. She recently transferred to Crestmont because it is one of the schools in the district that will be piloting curriculum units based on the new Common Core State Standards (National Governors Association Center for Best Practices and Council of Chief State School Officers 2010). In addition, Ms. Carlson believes in the added value of Crestmont's diverse student population. Crestmont is more ethnically and socioeconomically diverse than her previous middle school, Diablo Valley. Although Diablo Valley is a high-performing middle school, according to achievement test scores, Ms. Carlson was not as satisfied with the school's scores as her colleagues. She knew that many students had not developed the conceptual understanding to match their algorithmic mastery, and there were glaring disparities for some students. But whenever she voiced this concern to her colleagues, they told her she was "creating issues."

Ms. Carlson was ready for a change. At Crestmont, she teaches Baye and twenty-two other sixth graders in a "support" math class designed to address the mathematics needs of struggling students. Ms. Carlson has the following exchange with Baye after assigning a mathematics task to her students:

Ms. Carlson:	Why haven't you gotten started, Baye?
Baye:	[*Whispers under his breath*] Because this is stupid. The problem makes no sense. Who cares how many different sizes the rabbits' playpen can be.
Ms. Carlson:	You haven't even tried the problem yet.
Baye:	[*Looks briefly at the students sitting around him and mumbles*] What difference does it make? I can't do it. That's why I'm in this class, right?

The bell rings before Ms. Carlson can answer, and Baye packs up and rushes out the door. Ms. Carlson reviews the class papers but can't get Baye out of her mind. "What am I going to do?" she wonders. "How do I get him to try? He seemed embarrassed when I was talking with him. How am I going to motivate these kids to want to learn math? How am I going to get them to believe they can do math?"

Later, at home that evening, Baye has the following encounter with his father:

Father:	How was school today?
Baye:	[*Shrugs*] OK, I guess.
Father:	Did you finish your homework after school? [*Pauses while Baye is silent*] What's wrong?
Baye:	Nothing.

Baye throws his backpack on the table and heads outside. As his father takes the backpack off the table, he notices a crumpled piece of paper. It is a math assignment, but there is nothing on the page. He sighs and says to himself, "I thought this year would be different. He hates math. I don't know how to help. I didn't have any trouble with math in school. What am I going to do?"

From our experiences with students, teachers, parents, and caregivers, we know that situations like the one involving Baye arise every day. We know that many teachers reflect deeply on these kinds of encounters with students. They are concerned about students feeling discouraged and frustrated by mathematics. They want to change the students' negative views of mathematics and themselves as mathematics learners. We also know that many parents express similar concerns about their children in relation to mathematics. When their child is distressed, they want to help. In this case, Ms. Carlson and Baye's father are concerned and desire to help. They are at a loss.

In contrast, we have frequently heard educators make statements like the following about certain students, their families, and mathematics learning: "My students come from impoverished backgrounds. They can't handle that kind of math." We have also heard race-based and culture-based statements about Black and Latinx families "not valuing" their child's education. In contrast, white and Asian families supposedly "push their children" to do well in mathematics. Furthermore, some teachers question whether all children should study advanced levels of mathematics, echoing common public refrains like, "Not everybody needs to study algebra. They just need to know the basics," or "There should be other options for those who are less mathematically inclined," or "I didn't do so well in math, and I did just fine. I have a successful career."

Statements such as these are quite common yet may not accurately reflect students' identities, abilities, or interests. They are also not the most reflective, equity-oriented approaches teachers can take. Moreover, these statements, although well intentioned, may perpetuate negative stereotypes about what mathematics is, who can learn mathematics, who supports mathematics, and why students should or should not pursue their mathematics studies. In our view of equity-based practice, it is important to challenge these discourses and replace them with strengths-based and affirming perspectives.

Beyond Changing Demographics

A common approach to engaging teachers about issues of equity is to broadly cite changing demographics and the increasing racial, ethnic, and linguistic diversity in the student population. However, such references do not always lead to the kind of deep reflection about practice necessary to strengthen mathematics learning and positive mathematics (and other) identities students hold. Although many teachers may be aware of and receptive to broad, general discussions of equity and justice in mathematics, general awareness may not be enough to motivate them to address the particularities of working across and within particular student populations. In addition, focusing on demographic changes often positions students from different ethnic or language groups at odds with dominant student groups. The students seen as diversifying the student population (often identified in racial, ethnic, or class terms) are frequently called on to conform to the established standards or norms. In our view, meaningful inclusion and interactions with students necessitate knowledge of their personal, family, and community backgrounds and their social realities. Gaining this knowledge may require additional effort and knowledge on the part of teachers and administrators to fully meet the mathematical learning needs of their students.

For example, several years ago, one of the authors, Danny Martin, presented at a regional conference of the National Council of Teachers of Mathematics (NCTM) in Chicago. In his presentation, he argued for the relevance of identity as a key consideration in the mathematical experiences of Black children. In particular, Martin contended that teachers need to try to understand how these children make sense of what it means to be doers of mathematics, and, simultaneously, how they make sense of what it means to be Black, on the basis of their emerging understandings of their life experiences and social realities. He suggested that the emerging understandings developed by students reflect not only the assertions that students make about who they are but also the ways in which they accept or resist the racial and mathematical identities that are imposed on them by others, including teachers, peers, parents, caregivers, community members, and the media.

In his presentation, Martin encouraged the audience, especially teachers, to think more deeply about these identity and learning issues by considering two focused questions:

1. What does it mean to be a learner and doer of mathematics in the context of being Black?

2. What does it mean to be Black in the context of learning and doing mathematics?

Martin asked teachers to consider the range of responses that might emerge among their students, given their social realities, and how those responses might be useful to teachers as they reflect on their work with Black children.

At the conclusion of Martin's talk, a young white female teacher in the audience raised her hand. She began her comments by noting that she was a teacher in a Black school located in a Black community on Chicago's South Side. She said that although she had been teaching at the school for a few years, she had "never thought about what it means for my students to be Black." Martin's talk was a revelation to her as a teacher of Black children.

We contend that it is reasonable to ask why this presentation and the questions raised by Martin were such revelations to this teacher. How did this teacher's inattention to the power and relevance of Black identity in the lives of her students evolve? And to what extent had her views affected her mathematics instruction in classrooms with Black children up to this point? Although one could debate Martin's focus on Black racial identity and its role in mathematics learning and teaching, attention to the local and larger contexts of this teacher's school and the children she works with daily highlight its relevance. For example, the public school system in Chicago is the third largest in the United States, with Black children making up nearly 40 percent of the student population in the district. Chicago's public schools are racially segregated, mirroring the racial separateness of the city's neighborhoods. The average Black child attending a school in the district is in a school that is more than 80 percent Black. We contend that race and racial identity are highly salient in such contexts. Our reaction to this teacher's response to Martin's presentation does not imply that we believe that the students *being Black* should have been the *only* consideration in the teacher's interaction with her Black students. Martin also argued for teachers of mathematics to move beyond demographic data and reflect on those identities that might be most salient and important to students and to understand how these identities might shape student and teacher engagement with school mathematics.

In a more recent conference presentation focused on affirming the ordinary brilliance of Black children in mathematics (c.f. Leonard and Martin 2013; Wilkes 2022), Martin was met with strong resistance from a university colleague who could not accept Black children's brilliance as axiomatic. The mathematician colleague suggested that he had only encountered two or three brilliant people in his lifetime and wanted more evidence and proof for the brilliance being claimed about Black children. The subsequent exchanges between Martin and his colleague became more and more animated and heated. From our points of view, we see the objections of the faculty member as a form of refusal, an explicit unwillingness to juxtapose Black identity and brilliance.

Although Martin focused on Black children, all students bring both school and life experiences to the classroom, and these have an impact not only on how they perceive themselves as mathematics learners but also on how others see them. A goal for all teachers should be to learn enough about these experiences to engage, support, and teach all students, whether they are Black students from an urban context, new immigrant students in a rural town, or affluent students in a private school. We call on teachers to see Black children, and others, in expansive ways, including their ordinary brilliance.

We call on teachers to see Black children, and others, in expansive ways, including their ordinary brilliance.

Throughout this book, we suggest that a number of identity-related issues can emerge as being relevant to how teachers support mathematics learning of their students in different contexts. How teachers recognize and respond to these issues will have an impact on how they address the questions raised at the beginning of this chapter.

Rethinking Equity: Toward Humanizing and Liberatory Mathematics Education

In the previous edition of this book, our goal was to help teachers develop equity-oriented practices in relation to mathematics. We embraced a perspective on equity that supports teaching practices and reflective tools focused on the empowerment of the whole child. As a result, this equity-based approach included attending to the multiple identities—racial, ethnic, cultural, linguistic, gender, mathematical, and so on—that students develop and draw on as they learn and do mathematics. In support of that holistic view of equity, we offered the following description of what we believe teachers owe to all students:

> All students, *in light of their humanity*—their personal experiences, backgrounds, histories, languages, identities, and physical and emotional well-being—must have the opportunity and support to learn rich mathematics that fosters meaning making, empowers decision making, and critiques, challenges, and transforms inequities and injustices. Equity does not mean that every student should receive identical instruction. Equity demands responsive instruction that promotes equitable access, attainment, and advancement in mathematics education for each student.

We firmly believe that this perspective on equity challenges common notions that students need to learn math "in spite of" or "regardless of" who they are. We argue that students need to learn mathematics *in light of* who they are and the diverse gifts that they bring to their experiences every day. In the case of Danny Martin and the young teacher, this more holistic view of equity-based teaching practice would require attending to and understanding Black children's emerging and developing racial identities in the context of local and larger social realities in which they live every day. In the case of those who act as the arbiters of ability and competence, like his university colleague, it would entail accepting the ordinary brilliance of Black children (Wilkes 2022). In the case of Baye, it would mean understanding how all the different experiences in Baye's school life, home life, faith life, athletic life, and cultural life may affect the ways that he experiences school mathematics and that he, his father, and his teacher see him as a mathematics learner. Furthermore, this equity perspective demands attention to the ways that societal views of mathematics performance may fuel stereotypes (for example, the notion that Asian students are good at math) and obstruct the development of a positive mathematics identity.

We also recognize that this holistic view means embracing life complexities that may support and *challenge* children to learn mathematics and develop their mathematics identities. Teachers can inspire students beyond or apart from difficult life circumstances, and they can take advantage of the strengths that all children bring to school. They also can disrupt or eliminate, rather than perpetuate, negative images of what it means to learn mathematics and beliefs about who can learn mathematics. They can develop strong partnerships with parents and caregivers to support a child's learning of mathematics.

In reflecting on the version of equity offered in the previous edition, we realize that it was incomplete. It was a work in progress. Our perspective on equity in the first edition made less apparent several considerations that we now make explicit.

First, after reviewing equity-oriented and related efforts focused on diversity and inclusion over several decades and living through the realities of recent societal events like the murders of George Floyd, Breonna Taylor, and other Black Americans; seeing news reports of migrant families and children separated and locked up in cages on the southern borders of the United States; experiencing the disproportionate impact of COVID-19 on Black, Latinx, Native American/Indigenous, and Asian American communities; and the continued educational tinkering under the guise of reform, we realize that equity is not a sufficient counterbalance to deep structural forces like white supremacy, antiblackness, and xenophobia whose reach can extend to and be supported by mathematics education. Equity in everyday practices of the classroom does not shield students from these structural forces and the inequities, oppression, and hierarchies they produce. White supremacy, antiblackness xenophobia, and other oppressive systems are adaptive and often self-correcting (Davis and Jett 2019; Martin 2019; Martin, Price, and Moore 2019). These adaptations allow for incremental changes and the appearance of progress (e.g., equity) while the fundamental oppressive character of these structural forces on mathematics education remains unchanged for many students. In other words, equity often is achieved within the system as it is, not as it should be. We discuss this shift in our thinking in Chapter 4, where we take up standards and standards-based practices as political entities used to support equity.

Second, in the previous edition, we may have given the impression that a positive mathematics identity was the ultimate culmination of mathematics teaching and learning. Of course, we would like all students to develop such identities. However, we are well aware that positive math identities do not protect students from white supremacy, antiblackness, and xenophobia. A Black girl who is six years old may be in the early stages of developing a positive math identity, but this may not shield her from being arrested, handcuffed, and processed like a criminal because she had a tantrum at school. This was the case for six-year-old Kaia Rolle in September 2019 when she was arrested, handcuffed, and led out of the Lucious and Emma Nixon Academy in Orlando, Florida. According to her grandmother, Merilyn Kirkland, Kaia had been identified as "gifted" in mathematics by the state of Florida.

Third, rather than stop at equity and positive mathematics identities, we extend our push to fostering humanizing and liberatory forms of mathematics education that are worthy of children and teachers (Davis 2021; Goffney and Gutiérrez 2018; Leonard 2020; Yeh and Otis 2019). A humanizing and liberatory approach requires explicit attention to white supremacy, antiblackness, and xenophobia and dismantling structures and practices that support them. Teachers, students, parents, caregivers, and community members should be able to live and learn unfettered by these forces with no inclination or incentive to reproduce them in any form. Humanizing and liberatory forms of mathematics education resist mathematics education being appropriated to reproduce social hierarchies. They resist all forms of experiential, symbolic, and systemic violence (Martin, Price, and Moore 2019). This includes deficit-based thinking in current schooling policies and practices that stubbornly persist and dehumanize children from being fully free to engage with and learn mathematics (see Chapter 5).

Fourth, we provide elaborations on what we mean by *meaningful* and *empowering* forms of mathematics (see Chapter 4). Prompting these elaborations is a shift from considering which students are worthy of privileged forms of mathematics education to whether particular forms of mathematics education are worthy of students.

Conclusion

This book offers guidance and support to rethink instructional practice and embrace a humanizing and liberatory approach to mathematics education. Enriching one's practice in this way takes conviction and courage. Reflecting on these possibilities and their impact on instructional practices is key. We encourage teachers to take *action* to change instructional practices and policies in ways that can strengthen student learning and cultivate empowerment, particularly in children who continue to be marginalized.

DISCUSSION QUESTIONS

1. Why is mathematics important for students to learn? Whose interests are served by the reasons that you give?

2. If you could consult with Ms. Carlson and Baye's father, what would you discuss about how best to support Baye as a math learner? Reflect on why those discussion points are important in relation to mathematics learning and identity.

3. Reflect on Danny Martin's viewpoint that it is important for teachers of mathematics to consider the interaction of racial identity and mathematics identity in children's experiences of learning school mathematics. Do you have questions about this perspective?

4. After reading this chapter and reflecting on the examples in it, what strategies have you used to learn about the school and life experiences of students in your classroom?

5. What questions do you have about the impact of students' math identities on their learning and the structural forces that shape them (e.g., white supremacy, antiblackness, and xenophobia)?

6. How do you interpret the shift from equity to humanizing and liberatory forms of mathematics education? What do these concepts mean to you? What are the implications of this shift for your teaching practices?

Chapter 2

Identities, Agency, and Mathematical Proficiency: What Teachers Need to Know to Support Student Learning and Empowerment

Consider the comments of Terrell, a fourteen-year-old boy, about himself:

> Ever since I started playing hockey, my dad has been on my case to do my work. Keep my grades up in school. If I don't, that's the end of my hockey career. The most respect I get is from hockey.... I want to be the best in hockey, so I work hard in school to be able to play hockey.... People think of hockey as a white man's sport. But I think if a man wants to play hockey or a man wants to do something he wants to do, then he should be able to do it without anybody questioning how he does it or why he does it.

Terrell was one of thirty-five students interviewed as part of a larger research study that focused on academic and mathematics success and failure among African American middle school students (Martin 2000). Analysis of this excerpt, and the longer interview from which it came, showed that Terrell's strong identity as a hockey player served to strengthen his identity as a good mathematics student and helped to keep him among the highest achievers in his school.

A central goal of this chapter is to discuss the inextricable links between mathematics learning and identity. Specifically, we focus on some of the many forces that shape students' *mathematics identities*—how students see themselves and how they are seen by others, including teachers, parents, caregivers, and peers, as doers of mathematics. We also give attention to other identities that students develop in and out of school and discuss how those identities shape and are shaped by students' life experiences. Then we discuss the concepts of *mathematical agency* and *mathematical proficiency* to highlight important behaviors, dispositions, and skills that teachers can help students strengthen to support positive mathematics identities. We conclude the chapter with a discussion of how teachers can reframe the identities that they ascribe to students—identities that might be based on stereotypes or limited knowledge of student backgrounds—to help strengthen mathematics teaching and learning in the classroom.

What Is Identity, and Why Should Teachers Be Concerned about It?

Identities can be defined as "the ways that people come to conceptualize themselves and others" and how they act because of those understandings (Cornell and Hartmann 1998, p. xvii). Identities can emerge in the form of *stories* that announce to the world who we think we are, who we want to become, or who we are not.

Student identities are diverse and complex. They can be faith-based (strong Muslim or Christian identities, perhaps) and family-based (identities as "good sons" or "good daughters," for instance). Identities of young people can also include early identifications with careers as doctors, lawyers, teachers, engineers, firefighters, train conductors, artists, or sports professionals, for example. These identities are important; they can serve as sources of strength and motivation to do well in school, in general, and in mathematics, in particular (Martin 2000).

We believe that children's developing identities should be important considerations in the daily work of all teachers. Teaching involves not only developing important skills and conceptual understanding in mathematics but also supporting students coming to see themselves as legitimate and powerful doers of mathematics. This understanding of children's identities, especially in relation to mathematics, can give teachers a better understanding of how and why some students make positive connections with mathematics and others do not. With this enhanced understanding, teachers can adjust their practice to support and strengthen a child's learning of mathematics and his or her persistence as a confident mathematical learner.

What Are Mathematics Identities?

We define *mathematics identity* as the dispositions and deeply held beliefs that students develop about their ability to participate and perform effectively in mathematical contexts and to use mathematics in powerful ways across the contexts of their lives. Depending on the context, a mathematics identity may reflect a sense of oneself as a competent performer who is able to do mathematics or as the kind of person who is unable to do mathematics.

Mathematics identities can be expressed in story form. These stories reflect not only what we say and believe about ourselves as mathematics learners but also how others see us in relation to mathematics. Teachers, peers, parents, and caregivers can all exert an influence on the mathematics identities that students develop. A key consideration about mathematics identities is that they are strongly connected with the other identities that students construct and view as important in their lives, including their racial, gender, language, cultural, ethnic, family, faith, and academic identities.

For example, Berry (2008) interviewed and observed six African American middle school boys who were able to reflect on their mathematical experiences and how those experiences shaped their mathematics identities. Although Berry was interested in how these boys learned and saw themselves in relation to *mathematics*, he did not minimize attention to their *racial* identities as African Americans, because both identities were salient in the school experiences of these boys. One of those interviewed was Cordell, whose narrative provides a glimpse of his emerging mathematics, academic, "good son," and African American identities and shows how his experiences in and out of school wove those identities together.

My name is Cordell, and I am an eighth-grade student at Memorial Middle School. I am an only child, and I live with my mother. I know that my mother, being a single parent, has a tough job, so I have had to take on more responsibilities than other kids do, and I have learned to be independent. My grandmother and aunts help my mother by encouraging me to make good decisions and make sure that I stay on the right track. My grandmother and mother talk to me about doing well in school and make sure I do my work. My mother is always saying I better do well in school if I plan on going to college.

Math is my favorite subject because it is my easiest subject. Math is interesting and fun because in math you have to think and keep trying until you get it right. I was first drawn to math in the third grade when we started to learn how to multiply.

I knew I was good because I learned to multiply earlier than the other kids in my class. I am glad that I was good at math at a young age, because that put me ahead of the other kids in my class. My third-grade teacher divided the class into groups, and I was with the group that got the harder problems. This made me feel like I was smart.

When I was in fourth grade, I started getting into trouble because I was bored with school. My teacher was teaching me things I already knew, so I would start playing around in class. My mother thought I was not being challenged enough and that is why I got into trouble. After a few conferences with the teacher and the principal, my mother felt that I should be tested for the AG [academically gifted] program.

The teacher and principal did not want me tested because they felt I was not gifted. My mother thinks the reason they did not want to test me was because I am Black. She stayed on the teachers and principals until I was tested. I did well enough to be placed in the AG program midway through my fourth-grade year.

Cordell's narrative reveals a number of identities that are important to him and that are interwoven in his sense of self as a mathematics student. These identities include being a middle school student, an only child, an independent good son, a self-acknowledged smart student, and a Black boy who some school officials think is not gifted in mathematics. Cordell's narrative also helps to demonstrate our claim that students negotiate a number of complex identities that emerge as important to them. These identities can find support from parents, caregivers, and teachers, and other significant people in students' lives.

As Cordell's narrative helps to demonstrate, parents, caregivers, and teachers can have profound influences on their children's mathematics identities in response to the messages they send about their competencies and abilities. These messages can emerge in the stories that children tell about their mathematical experiences.

It is important to note that mathematics identities emerge not only through the stories that students tell and that are told about them but also through the *behaviors* that they demonstrate to help position themselves as certain kinds of people (for example, good math students) or as members of a particular group (high achievers, for instance). A student's correct and confident use of mathematical language and argumentation strategies, supported by positive feedback from teachers and peers, could help to reflect or shape a positive mathematical identity. These identity-affirming (or identity-challenging) behaviors can influence the kinds of learning experiences and social relations that students have with others (Cornell and Hartman 1998). Students who have been identified and behave as "gifted" mathematics students among their peers may dominate classroom interactions and activities in an attempt to maintain their status. Students who believe that they are not good at mathematics may remain silent in small-group interactions because they fear that other students will judge them. Language-intensive practices that demand increased levels of math discourse may come to favor or privilege some students (native English speakers, students who are outspoken) over others (English language learners, shy students), allowing the former to assume leadership roles, elevate their status as doers of mathematics, and improve their mathematics communication skills.

Similarly, classroom activities that reward speed as ideal mathematics behavior may lead students to believe that being "good at mathematics" means being able to recite multiplication facts or carry out calculations quickly. Students who are more deliberate in their work may see themselves as being not good at mathematics. Moreover, as criteria emerge to establish who gets labeled as "smart" or "gifted" or "slow" or "proficient" or "at-risk," students will come to see themselves in particular ways relative to other members of their mathematical communities. Instead of becoming more valued members of their classroom communities, they may come to see themselves as outsiders.

Thus, many influences shape a student's mathematical identity—some negative and some positive. It is important for teachers to understand the impact of the instructional decisions that they make, and the social and academic norms that they create, on a child's mathematics identity.

Mathematical agency

The definition of *mathematics identity* presented earlier in this chapter includes "the ability to participate and perform effectively in mathematical contexts." This behavioral aspect of mathematics identity can also be captured by the term *agency*. Several mathematics educators have taken up the idea of *mathematical agency* and documented it among students and teachers in classroom settings. Turner (2003), for example, has drawn on her work with Mexican American and Mexican children in the southwestern United States to conceptualize *critical mathematical agency* as students' capacity to "identify themselves as powerful mathematical thinkers who construct rigorous mathematical understandings, and who participate in mathematics in personally and socially meaningful ways" (p. iv). Gresalfi and colleagues (2009) characterized agency with respect to opportunities to complete mathematical tasks, and they distinguished two forms of mathematical agency: *disciplinary agency* and *conceptual agency*:

> Recalling facts or definitions and executing procedures involve disciplinary agency; there are correct answers, and a student either gets it right or doesn't. Procedures with connections and, especially, doing mathematics generally involve conceptual agency, with students being positioned to take initiative in constructing meaning and understanding of the methods and concepts that are the subjects of their learning. (p. 56)

Both Turner's and Gresalfi and colleagues' conceptualizations of agency help to highlight that students are active participants in, rather than passive recipients of, their mathematics education experiences. They can exercise these forms of agency in productive ways—resisting negative identities that are imposed on them, developing mathematical strategies within the context of small-group work, or using mathematics as a tool to understand their life circumstances or events in the world. Creating opportunities for students—particularly those who traditionally have had less access to powerful mathematics and mathematical practices—to engage in productive forms of agency should be a goal for all teachers.

The idea of mathematical agency is not confined to individual students. Classrooms of students can exhibit *collective mathematical agency* when teachers and their students act together to solve problems, working from the shared belief that viable strategies can be developed, and solutions can be found. Different students can contribute different elements to this collective agency. Some students might contribute productive reasoning strategies. Other students might make computational contributions. Others might contribute through whole-class explanations of particular mathematical concepts or by asking questions that help to clarify problems and concepts for themselves and their classmates. Teachers can also encourage students to assume various roles that provide them with opportunities to make viable contributions to classroom activities and practices. Some students with bilingual competencies might be assigned roles as translators for their peers whose first language is not English so that these students will not be left behind. Teachers can further contribute to this collective agency by helping to establish classroom norms and rules for behavior that encourage cooperation and risk-taking during problem solving rather than strict competition (Featherstone et al. 2011; Horn 2012).

Reflective teacher practice that is committed to equity will include the development of tasks, activities, and classroom cultures that encourage students to exercise their positive mathematical agency, individually and collectively. These forms of agency can contribute to students' developing positive identity-related stories and behaviors that affirm and demonstrate these identities.

Mathematical proficiency

Creating these expanded opportunities for students to learn mathematics and develop productive mathematics identities with powerful agency will also require teachers to develop a broader concept of what counts as mathematics proficiency. As outlined in *Adding It Up: Helping Children Learn Mathematics* (National Research Council 2001a), teaching for mathematical proficiency no longer should include a singular focus on having students develop computation skills and memorize algorithms, perhaps privileging those students who believe mathematics is about doing computations quickly. As the book suggests (see figure 2.1), mathematical proficiency should include developing *conceptual understanding* (comprehension of mathematical concepts, operations, and relations), *procedural fluency* (skill in carrying out procedures flexibly, accurately, efficiently, and appropriately), *strategic competence* (the ability to formulate, represent, and solve mathematical problems), *adaptive reasoning* (the capacity for logical thought, reflection, explanation, and justification), and *productive disposition* (a habitual inclination to see mathematics as sensible, useful, and worthwhile, coupled with a belief in diligence and one's own efficacy).

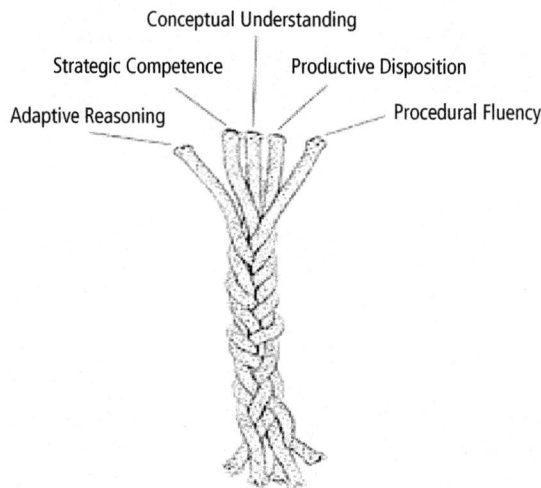

Fig. 2.1. Five strands of mathematical proficiency
(National Research Council 2001a, p. 5; reprinted by permission)

Although the five strands of mathematical proficiency outlined in *Adding It Up* are intertwined, as the figure suggests, one or more may emerge earlier in some students than in others. Some students might demonstrate deep conceptual understanding but might not yet demonstrate strong strategic competence. Other students might demonstrate strong procedural fluency but might not yet display a productive disposition toward certain kinds of tasks and practices. Teachers will need to reflect on which classroom and curricular practices provide the best opportunities for these components of mathematical proficiency to emerge.

Connections to these strands of mathematics proficiency can also be found in the mathematical practices identified and emphasized in the Common Core State Standards for Mathematics (National Governors Association Center for Best Practices and Council of Chief State School Officers 2010) as well as process standards in state frameworks such as the Texas Essential Knowledge and Skills for Mathematics K–12 (Texas Education Agency 2012). The Standards for Mathematical Practice "describe ways in which developing student practitioners of the discipline of mathematics increasingly ought to engage with the subject matter as they grow in mathematical maturity and expertise" (p. 8):

1. Make sense of problems and persevere in solving them.
2. Reason abstractly and quantitatively.
3. Construct viable arguments and critique the reasoning of others.
4. Model with mathematics.
5. Use appropriate tools strategically.
6. Attend to precision.
7. Look for and make use of structure.
8. Look for and express regularity in repeated reasoning.

We believe that these shifts in characterizing mathematical proficiency can help to foster reflective and equitable mathematics practice among teachers *and* the development of positive identifications with mathematics among students. Instructionally, these shifts offer teachers an increased number of pathways and points of entry to assess students' mathematical development. For students, these broader conceptions of mathematical proficiency provide multiple ways to demonstrate their competence. That is, these expanded possibilities for developing and demonstrating mathematics competence can represent substantial opportunities to learn and engage in mathematics. We believe that *how* students experience mathematics in their classrooms shapes their views of mathematics and themselves as mathematics learners and doers. Thus, how mathematical proficiency is defined and communicated to students has a powerful impact on their mathematics identities and their exercise of various forms of agency.

Seeing the Multiple Identities of Students

Although we give primacy to mathematics, we emphasize the importance of recognizing the range of identities beyond mathematics that students spend their time and energy developing or that others may assign to them. Acknowledging these identities can lead to richer, more meaningful understanding of children and their lives. Terrell's identities as a hockey player and a good student, for example, serve to challenge popular societal and school-based perceptions and negative stereotypes about what it means to be a young African American male. It is important to note that although Terrell was able to defy negative stereotypes by maintaining high grades in mathematics, expressing positive attitudes toward mathematics, and demonstrating classroom behaviors typically associated with being a good mathematics student, he did not identify mathematics as his favorite subject or use mathematics as the *primary* source for constructing his overall academic identity. In fact, English was Terrell's favorite school subject:

> I like English because I like to write. I'm not much of a poem writer but I want to write poems. That's what I want to do. Nobody knows. I never told anybody that I want to write poems, but that's what I want to do. So I like English. My mentor is Langston Hughes. I look up to him. (Martin 2000, pp. 146–147)

These remarks by Terrell serve to remind us that, unless asked, students may be unwilling to reveal important aspects of their many developing identities. Similarly, students can be very adept at invoking particular identities to serve particular needs in both school and nonschool contexts.

It is equally important to stress the significance of identity-based considerations for students whose identities are often taken for granted. In many school-based discussions and policies, the categories "Asian" and "Latinx," for example, are often associated with beliefs about students who do well in school, in the case of the former category, and with children who struggle and fail, in the case of the latter category. Often these associations exist with no consideration of the varied experiences of different subgroups and the varied home and cultural experiences and roles of children within those subgroups.

Orellana (2009) highlights the family roles of immigrant children and the complex identities they take on in relation to both their family and school lives. She offers a brief profile of a fourteen-year-old girl, Cindy, and describes her role as family translator. Orellana explains that Cindy liked this role because "she learned more about other people in her family and about herself…; it made her feel smart; she learned more words in her two languages, English and Chinese" (p. 9). In describing how she believed these experiences distinguished her from her peers, Cindy was simultaneously able to reveal the out-of-school mathematical needs (for example, budgeting and making banking transactions) that she was meeting to ensure her family's well-being:

Sometimes I think I invaded people's privacy, like they have to tell me over the phone, like deposit statements and stuff like that. I know exactly the house's wages and stuff like that, and I tell my parents, and they don't really care. I just know, and I translate it. While the other kids, they ask for things, I'm not trodding down people of my own age, but some people they just ask for things, like "Can I have a bike, can I go swimming, can I go to summer camp, can I have a new pair of Nikes?"… Their parents keep saying, "Do you know how hard I work for the money to pay bills?" They don't know exactly how much is in their bank deposits, the bills and stuff. But *I* know personally because I write the bills. I write the checks. (Orellana 2009, p. 9)

In Cindy's case, her family roles and out-of-school experiences helped to shape her identity as a daughter, a translator, and a budget manager. Her identities intersect with one another in ways that reveal complexities that nonimmigrant children may not experience.

Our point in presenting the preceding examples is to stress that *all* students develop, resist, and try on many different identities as they make sense of their experiences in school and nonschool contexts. These identities, in our view, are important in shaping how students come to see themselves as mathematics learners.

Reframing Identities to Strengthen Mathematics Learning

We encourage teachers to take an active role in shaping positive mathematics identities among their students by also attending to the other identities that students are developing. But to succeed, teachers must do so in ways that move away from negative stereotypes (such as identifying Black males as thugs or identifying Latinx and Asian children as "illegal aliens" or "anchor babies") and avoid reducing children to nameless, objectified data points such as "bubble kids" or "bottom quartile kids" in larger conversations about assessment. Increased awareness and understanding of students' multiple and complex identities may require that teachers move beyond the simple categories that seem to work for sorting and ranking purposes. Demographic and socioeconomic labels such as "Black," "white," "Latinx," "Asian," "Native American," "immigrant," "urban," "rural," "middle class," "gifted," "limited English proficient," "multilingual," "at-risk," "disabled," and "poor" are common in school discourse but are often used in ways that mask their complexity and intersection.

We ask teachers to reflect on how their students might make sense of these labels and categories. Understanding how and why students come to resist or accept particular labels can be helpful in understanding their engagement or lack of engagement in mathematics. For example, do students internalize positive or negative stereotypes about mathematical intelligence and ability that are associated with their own or others' racial and ethnic identities? In the context of discussions of racial achievement gaps, do students who identify themselves as "Native American," "African American," or "Latinx" see themselves as intellectually inferior to students who are identified as "white" or "Asian"? What effects do discussions about "illegal" immigration have on the academic engagement of undocumented students or students with undocumented family members? How do students from various Asian subgroups respond to stereotypes about mathematical superiority or the expectations that come with being labeled as a "model minority"?

As teachers reflect on how they might simultaneously support and affirm students' racial, gender, cultural, ethnic, academic, and mathematics identities, we encourage them to seek ways to reframe negative views of the identities that reflect limited knowledge of students' background experiences and social realities. We offer some examples and suggestions in the discussion that follows.

From acceptance of, to resistance to, "model minority" myths

Societal and educational discourse often holds up students who are identified as "Asian," "Southeast Asian," and "East Indian" as "model minorities" in comparison with students who are identified as "African American," "Latinx," or "Native American." Such discourse communicates the idea that Asian students are academically superior, come from cultures and families that value education, and have successfully assimilated into American society. Because Asian American students have minority status and are perceived as having overcome language and cultural barriers to achieve their success, the belief is widespread that students from other minority groups should be able to overcome their circumstances and achieve at much higher levels. Instead of assuming that African American, Latinx, and Native American students come from racial and cultural backgrounds where education is valued, the dominant narratives suggest that "cultural deficits" are the cause of underachievement by these children.

Several Asian American scholars, including Lee (2005, 2009), Louie (2004), and Shah (2019), for example, have challenged the idea of the Asian model minority and have pointed out that this characterization of schooling experiences often overlooks the educational struggles of various Asian American subgroups, including Laotian, Hmong, Mien, and Cambodian students. This myth also overlooks the experiences of poor and working-class Chinese and Vietnamese students, for example, who may not be high achievers but may drop out of school or be on the verge of doing so. As a result of the myth, these students are not likely to receive the support services that they need to improve their academic standing or help them remain in school.

We suggest that teachers reflect critically on this myth and understand that Asian American students are varied and diverse in their identities and backgrounds. Furthermore, their mathematics experiences, like their experiences in broader societal contexts, reflect the impact of issues of race, class, gender, and culture.

From limited English to multilingual language brokers

Many states and cities are experiencing changes in demographics that often bring dozens of languages into their schools and classrooms. When children from culturally and linguistically diverse backgrounds enter school, their home language and language status often become primary markers for their identities. How teachers respond to these language identities is important. We know, for example, that these students are typically referred to as "limited English proficient," with *limited* as the operative word. Although this assignment of identity to these children is accurate to a certain degree, it is itself limited. These students are in the process of becoming bilingual or multilingual, with English becoming for them a second, or even a third, language.

The label "limited English" masks the multilingual backgrounds and experiences of many students. The reality is that many of these children are positioned within their families as *language brokers* who must navigate nonschool contexts on behalf of their parents, despite their age and evolving development as English speakers. As we noted earlier in the case of Cindy's story, researchers have documented these brokering practices among students and families who have immigrated from Mexico, Central America, China, Hong Kong, and Korea. Orellana (2009) offers useful descriptions of these children's roles as language brokers:

> Children serve as language brokers because their families need their skills in order to accomplish the tasks of everyday life in their new linguistic and cultural context. Many teachers also need these children's skills. (p. 2)

> Language brokering involves activities in which children, often taking the lead with adults, facilitate their parents' abilities to accomplish what these adults would not be able to accomplish on their own. In the process, children also support their parents' acquisition of English language and literacy skills. (p. 104)

Multilingual children also serve as language brokers in the classroom by performing bilingual translations and mediations of oral and written texts from their first language to English and vice versa. They act in this role for their peers as well as their teachers (Manyak 2004). In addition, recent research has demonstrated the connection of language-brokering practices with increased levels of academic performance (Dorner, Orellana, and Li Grining 2007).

In the case of mathematics classrooms, we would argue that bilingual students at various levels of English proficiency can spontaneously find themselves serving as, or being positioned by others to serve as, language brokers during classroom interactions. More importantly, these students can bring to mathematics learning considerable strengths that may go unrecognized if the instructional focus is only on the use of English vocabulary and pronunciation rather than on additional ways in which these students communicate their ideas through gestures, representations, and their first languages.

For example, Moschkovich (2002) highlighted the ways in which Latinx bilingual students used an array of resources that supported their own and their peers' learning. During a middle school math class, students constructed rectangles of the same area (36 square units) and different perimeters while looking for patterns that related the perimeter to the dimensions of the rectangles. During a small-group discussion, one group of Latinas spoke primarily in Spanish while attempting to solve the problem. They struggled to come up with the Spanish word for *rectangle*, using other words, such as *ángulo* ("angle"), *triángulo* ("triangle"), and *rángulo* ("rangle") in their problem-solving efforts. Later, the teacher asked the small groups to present their ideas about mathematical relationships between the perimeter and the dimensions of the rectangle, and one of the students in the small group, Alicia, responded:

The longer the, ah … the longer [traces the shape of a long rectangle with her hands several times] the, ah … the longer the rángulo [rangle], you know, the more the perimeter, the higher the perimeter is. (p. 201)

What is important to note is that as the group's spokesperson, Alicia was serving in a role as a language broker for her group, for her peers, and for her teacher. She communicated a mathematical idea in English that was developed in her small-group discussions that occurred primarily in Spanish. Although she did not use the correct term, *rectangle*, in the explanation, the way in which she used gestures and mathematical objects, such as drawings of rectangles, conveyed the group's collective understanding of a relationship between the shape of the rectangle (with longer lengths) and the perimeter. Moschkovich (2002) notes the completeness of the explanation: "Although Alicia was missing crucial vocabulary, she did appropriately (in the right place, at the right time, and in the right way) use a construction commonly used in mathematical communities to describe patterns, make comparisons, and describe direct variation: 'The longer the _____, the more (higher) the _____'" (p. 203).

Focusing only on the limited use of correct English vocabulary and pronunciation rather than on the mathematical ideas and language that Alicia did communicate could negatively affect Alicia's and her group's views of their mathematical competence. Shifting to a language-broker perspective enhances teachers' opportunities to recognize the multiple resources and responsibilities that bilingual learners bring to mathematics learning and participation.

From "at-risk" to "resilient"

Quite often, the discourse about students who come from backgrounds that are not middle class or wealthy is characterized by negative assumptions about their skills, abilities, competencies, and motivation. For example, if students are identified as "poor," assumptions attached to this label might include the belief that little teaching or learning occurs in their homes and communities. In our experiences in school contexts and discussions with colleagues, we have frequently encountered opinions like the following: "These children come from bad neighborhoods. Their parents don't care. They don't value education. They have too many hurdles in their lives to focus on learning. These children can't learn algebra." These attitudes represent very limited conceptions of children, their families, and their competencies.

Even if students come from backgrounds characterized by poverty and limited resources, they often exhibit high levels of resilience and mathematical excellence in the face of these circumstances (Martin 2000). For example, Turner and Celedón-Pattichis (2011) analyzed the problem-solving competencies of Latinx bilingual kindergartners. These students exhibited mathematical excellence in solving increasingly complex word problems and showcased various strategies for correctly solving problems involving multiplication and division. These children from working-class immigrant families in the southwest United States demonstrated mathematical success equal to that of wealthier students, according to assessments from an earlier study.

Studies like these continue to challenge commonly held deficit views of children from poor and working-class backgrounds, students of color, and multilingual learners as automatically "at-risk" by virtue of their racial, ethnic, and socioeconomic backgrounds and levels of English proficiency. We argue that a focus on learning rather than on labeling is critical. Furthermore, finding ways to build on this resilience, instead of focusing solely on the conditions that make such resilience necessary, should be primary goals for teachers.

Conclusion

In this chapter, we have encouraged teachers to engage in deeper reflection not only on the mathematics that they will teach but also on the multiple identities that emerge as important to their students and how those identities can shape and be shaped by mathematics learning and classroom engagement. We assert that if teachers plan to support the development of positive mathematics identities and multiple forms of mathematical proficiency and mathematical agency, they must develop a deeper understanding of these multiple identities and the social realities of their students. We encourage teachers to understand the productive identities that students are developing and reframe the negative identities in ways that move beyond stereotypes and simplicity.

DISCUSSION QUESTIONS

1. What range of mathematics identities is expressed and performed by your students? What actions do you take to positively affirm your students' mathematics identities? What are some ways that you might get students to share their emerging mathematics identities?

2. What are some of the various identities that your students express and perform through the stories that they narrate in your mathematics classroom? In what ways do these identities support or hinder the development of positive mathematics identities?

3. What are some of the ways that students demonstrate their mathematical agency in your classroom? How do you model positive mathematical agency and provide opportunities for students to demonstrate this?

4. What are some of the stereotypes and assumptions that emerge in your classroom about who can or cannot do mathematics? How do you and your students deal with these stereotypes and assumptions?

5. What additional family roles do your students take on that might contribute to their positive development in mathematics?

Chapter 3

Know Thyself: What Shapes Mathematics Teacher Identities?

In Chapter 2, we defined *identity* as the stories that people share about themselves and what they view as important to them: their understanding of their place in the world and their core beliefs. We can also perform an identity as one way to let others know the kind of person we claim to be. These stories and performances are multifaceted and dynamic and grow out of our experiences in multiple contexts—school, home, sports, traditions, media, careers, and so forth. Our identities represent negotiations between who we claim to be and how others identify and label us. Teachers' professional identities—which can be thought of as their "sense of self as well as their knowledge, beliefs, interests, dispositions, and orientations toward their work and change" (Drake, Spillane, and Hufferd-Ackles 2011, p. 2)—are also negotiated and shaped across experiences in a number of contexts, including university credential programs, professional development, district/state/federal policies, school relationships, political forces, and professional learning communities (Beijaard, Meijer, and Verloop 2004; Olsen 2011).

This chapter examines experiences and contexts that can contribute to the development of *mathematics teacher identity*—an identity that consists of knowledge and lived experiences, interweaving to inform teaching views, dispositions, and practices to help children learn mathematics (Drake, Spillane, and Hufferd-Ackles 2011; Gresalfi and Cobb 2011). In focusing on mathematics teacher identity, we acknowledge that it is just one of many disciplinary identities that teachers develop and that it can inform and shape their teaching practice. For example, some elementary teachers might be avid readers and passionate about helping students learn to read. This might lead them to integrate literacy strategies across different subjects to deepen students' skills and passion for reading. A middle school teacher who was formerly a wildlife biologist and who loves science might have chosen teaching as a second career to share her enthusiasm for science with young people. Another teacher might also be an artist, motivating him to find different ways to infuse art across the curriculum. In all these cases, a specific disciplinary passion can shape a teacher's professional identity and instructional practice.

This chapter discusses a range of personal and professional experiences that can contribute to the development of a mathematics teacher identity. We start with teachers' stories of their own mathematics learning experiences. These stories are math autobiographies of new teachers just entering the profession. The stories represent a wide array of experiences with mathematics in and out of school, including specific challenges and supports that shaped the teachers' mathematics learning, as well as ways that these experiences shaped their teaching vision for mathematics learning in their own classrooms. Furthermore, we discuss how those experiences were shaped, in turn, by race, class, gender, and language. By drawing on personal narratives from new teachers, we focus additional attention on elements that influence the development of mathematics teacher identity in its early stages and how they are perhaps somewhat disconnected from current views of mathematics teacher identity.

We also explore the unique role that the high status of mathematics as a subject in the K–12 curriculum plays in how teachers see themselves in relation to their work and their students. In many schools, mathematics is a centerpiece of a school improvement plan. In the face of calls to increase access to and the rigor of school mathematics experienced by students in our schools (for example, the Common Core State Standards for Mathematics and eighth-grade algebra mandates), these demands can lead some teachers to question how effective they can be in helping their students learn mathematics. Teachers may also develop beliefs that conflict with demands about what math topics should be taught to which students. Thus, a teacher's acceptance or rejection of such changes exposes important elements of their mathematics teacher identity. The chapter concludes with a series of discussion questions aimed to motivate reflection on mathematics teacher identity and how it shapes your practice in the classroom.

Teachers as Math Learners

The nature of the mathematics learning experiences that teachers had in their own schooling has a powerful influence on the mathematics teachers identity that they develop (Drake, Spillane, and Hufferd-Ackles et al. 2011). Because a mathematics teacher's identity is, at least partly, rooted in a teacher's experiences as a mathematics learner, we must explore how those experiences may have been shaped, in turn, by race, class, gender, and language. Herein, we discuss these interconnections by examining the mathematics autobiographies of six new teachers (all whose names have been changed).

Taken together, these stories help to provide an understanding of how teachers' identities as mathematics learners shape their mathematics teacher identities and how these mathematics teacher identities, in turn, influence the decisions and actions enacted in elementary and secondary mathematics classrooms.

Shawna Jamison: Middle school mathematics intervention teacher

Shawna is a new middle school mathematics teacher with a K–8 multiple subjects teaching credential and a special education endorsement. Her attitude toward math has evolved over time. Shawna grew up in a white, middle-class family. Until high school, her school peers came from similar racial and class backgrounds. Although her high school was more racially diverse, her peers in advanced mathematics courses tended to be primarily middle class and white, with some Asian American students. At home, her parents provided "many math learning opportunities," including baking (measuring ingredients, for instance), gardening (planting seeds in relation to sunlight and area or volume of soil), and softball (batting averages, earned-run averages, and so forth). In elementary school, Shawna loved to explore math concepts in class. She thrived on solving math problems on worksheets; math seemed "easy and enjoyable." During her teenage years, mathematics became less important. Although she pursued mathematics through high school calculus, she claimed that she "did not take responsibility" for her learning. When she attended the large public university in her state, her performance on the math placement test resulted in her having to retake precalculus. This placement led to a difficult time in her mathematics learning experience. The university precalculus class was huge, and she found herself unmotivated to attend class or seek help. She decided to withdraw from the class and change her major to psychology, which required very little math. Shawna successfully graduated with her bachelor's degree. She returned to school eight years later to pursue a course of study leading to a teaching career. However, she needed a mathematics class to enter the teaching credential program. She decided to retake precalculus at the local community college. With the help of a supportive and knowledgeable professor, peer collaboration, and her own perseverance, she rediscovered her fondness for mathematics.

> The class work was challenging but accessible. I was motivated and engaged. Before tests, I studied with other students in the class. While it was difficult to take the risk of being an engaged math student after all these years and math failures, the sense of pride I felt in my math achievement was priceless.

Shawna's evolved attitude toward mathematics and her own mathematics learning shaped her commitment to teaching mathematics at the middle school level, particularly to students who might struggle with mathematics:

> Watching the students struggle reminds me of myself as I struggled with math. I can relate to not wanting to try for fear of failure. While I grew up in a very different environment than most of these students, the same things that helped me get back on track with math will help them: feeling like a valuable contributor to a community of learners, curriculum that is challenging and appropriate for their level, and motivation to learn.

Clearly, Shawna empathized with mathematically struggling students because of her own experience as a math learner. She also perceived "environment" differences between her own childhood experience and those of most of the students she worked with in the math intervention courses.

Shawna's math schooling experiences were characterized by a lack of interaction with racially and ethnically diverse peers. However, her first assignment was teaching math intervention courses in a school that served large populations of Black, Latinx, and Vietnamese students. In addition, many of her students "lived in poverty." She viewed the similarities and differences between herself and her middle school students in a positive light. Although some teachers might have found these middle school students frustrating, given their academic and socioeconomic backgrounds, Shawna viewed working in the mathematics intervention courses as "an exciting opportunity to empower these students." She believed that what had worked for her in overcoming her mathematics struggles—encouragement to make valuable contributions, "challenging and appropriate" curriculum, and motivation to learn—could also work for her students. This positive attitude framed her vision and beliefs about what makes an effective mathematics teacher and helped to shape her instructional practice in her classroom.

Leslie Park: Third-grade elementary teacher

Leslie is a new third-grade teacher with a general K–8 multiple subjects teaching credential. Growing up, Leslie had negative feelings about mathematics because she associated it with academic intelligence. However, she found comfort in the fact that she believed mathematics always had a "right answer" and that memorizing formulas and following procedures was how to excel in mathematics. Leslie identified herself as Asian American. She grew up in a small rural town with a large agricultural industry. Leslie attended school with primarily white, middle-class, or wealthy peers. She described herself as an "oddball," not only because she was a member of a racial minority in her school (one of two Asian American students in the school) but also because she struggled with the model minority myth and stereotypes related to being Asian and excelling in mathematics.

> The truth is, I didn't just want to be good in math, I wanted to excel. This is because I associated intelligence with the ability to do math. Perhaps this is something that the society has ingrained in me but it was what I believed. And because I wasn't good in math, I felt like I was not smart. There is a stereotype of Asians being math geniuses. Being an Asian American myself, I sometimes felt embarrassed because I did not fit into the Asian mold.

In high school, Leslie had ups and downs in her mathematics success. She did quite well in ninth-grade algebra but proceeded to get a C in tenth-grade geometry. She remarked, "Shapes, angles, and theorems were all too complex and analytical." She also took precalculus and calculus in high school, working very hard with a tutor every day after school to obtain a grade of B. Yet, because of her self-imposed pressure to excel, she was not satisfied with such "mediocre grades." She did not fare much better in college mathematics. She started with the university's basic mathematics course involving algebra and some precalculus and "struggled through it to receive a 2.9." At that point, she decided to select a major that would "steer [her] far away from anything that involved numbers." She chose to major in English literature.

Because of her lived experiences as a struggling mathematics learner, her awareness of not conforming to the Asian American model minority myth, and her views of mathematics as procedural, with success tied to one right answer and serving as a proxy for intelligence, Leslie wanted something different for her own students.

> I do not want students to view math as a subject that is dry, formulaic, and something they have to do on their own. Rather, I want them to see it as a subject that

has many different approaches and is collaborative. In high school, I wanted so badly to be the "smart Asian girl" that I never even bothered asking my fellow classmates how they got their answers. There was no group work in math. I plan to use this personal experience that I had with math and turn it into something positive in my future classroom.

Leslie's desire to be the "smart Asian girl" in her class while struggling with mathematics demonstrates how powerful the impact of racialized and gendered experiences can be on mathematics identity. Leslie's mathematics identity deterred her from working collaboratively with her peers to learn mathematics. However, it was this mathematics learning experience that now shaped her developing mathematics teacher identity. As an elementary teacher, she wanted to build a teaching and learning environment that was positive compared with what she experienced as a math student. She was committed to creating an environment that focused on multiple approaches to mathematics problems and leveraged collaboration.

Michael Allen: Fifth-grade elementary teacher

Michael is a fifth-grade teacher at an urban elementary school. He is a product of the school district in which he currently teaches. He identified as a white male and grew up in a white, working-class neighborhood. He went to school with the neighborhood kids through middle school. Michael's earliest memories of mathematics included the emphasis in third grade on timed multiplication fact tests. Because he was competitive, he was disappointed about not being the fastest, yet he was also "relieved that I was not one of the slowest either." At home, he played games with his family, such as checkers, Monopoly, and various card games. Michael was interested in sports, including football and baseball, and he regularly read the sports page. Sports "brought numbers to life" for Michael:

It was vitally important to know how far out of first place [the] Cincinnati Reds or Philadelphia Phillies were. I would even calculate batting averages as soon as I realized that it was not magic or arbitrary. I credit playing football in the street with helping me learn the multiples of seven before anybody else in my third-grade class.

These mathematics experiences at home provided some support for mathematics in school. However, in sixth grade, Michael had a teacher who routinely berated students for not understanding "simple procedures." For example, she claimed that there was only one way to divide fractions and that "it was easy if you know your multiplication tables." Michael worked diligently to try to find alternative ways to divide fractions:

I had to prove her wrong. I added, subtracted, multiplied, and divided my little twelve-year-old brain out. I spent time before school and after working on formulas. I even did fractions at recess, trying to figure it out. I was sure that I would solve the elusive riddle of an alternative method to dividing fractions. I came up with page-long solutions for problems [like] $3/4 \div 1/2$, all incorrect, of course. I think it took me two whole days of diligent trial and error to realize that the longer I experimented, the longer and more complicated the formula became. I eventually came to the conclusion that if there was another way to solve a simple fraction division problem, it was so complicated that nobody would ever use it anyway.

Despite this situation, it was clear that Michael was comfortable with mathematics. In middle school, he found mathematics "relatively easy." However, things were different in high school. Although his high school had more racial diversity, the students in his mathematics classes were primarily white and middle class. His mathematics grades plummeted because he refused to do homework. He characterized himself as lazy, but he always managed to "skate by" because he did well on tests. In his junior year, his mathematics teacher gave him an F. This made him academically ineligible to play football for some local private colleges, but he did manage to earn good SAT scores, which allowed him to play football at a small public institution. Unfortunately, his partying and fighting resulted in his dismissal from college. He worked for a few years and then reenrolled in community college, where he "aced" microeconomics and macroeconomics. Armed with this success, he made plans to pursue an economics degree at the large public university in his state. Economics required calculus. He started with precalculus and "bombed it." He hated the "coldness of the numbers and mathematical ideas." He switched his major to history.

Michael graduated and worked in construction for fifteen years before pursuing a teaching career. He found it "ironic" that mathematics had chased him away from economics but he used it every day in his construction management job in numerous tasks, including measuring, estimating, budgeting, and modeling. Michael's experiences with mathematics outside of school through sports, games, and construction solidified for him the importance of making mathematics relevant and linked to real-world situations and experiences. He especially wants to capitalize on the interests and natural curiosity of his students:

With my students I will try to tap into their understanding of real-world mathematics applications and their natural curiosity as a way to help them create more math knowledge and skill for themselves.

Steve Smith: Middle school science and mathematics teacher

Steve teaches seventh- and eighth-grade science, computer programming, and mathematics at an urban middle school. Steve is a white male and grew up in a middle-class, white neighborhood. The demographics of his elementary, middle, and high schools were similar. Mathematics "came easy" for Steve in elementary school. His sixth-grade mathematics teacher recommended that he take the advanced seventh-grade pre-algebra mathematics class. Steve felt very honored by his teacher's confidence in his success. The seventh-grade math class proved challenging. Lacking self-confidence about being able to solve problems, Steve would try to hide behind his book. However, the teacher, in his usual way of encouraging participation and discussion in class, would regularly call on Steve to present solutions. Through hard work and persistence, and with the help of a patient teacher, Steve received an A in the course. His mathematics success continued in eighth-grade algebra, where he received one of the highest marks in the course.

However, in ninth grade, Steve transferred to a prestigious private school far away from his friends. He hit a wall in ninth-grade geometry. He could not replicate the proofs demonstrated by the teacher. He felt alienated from his classmates because he was not an athlete or a "favorite" of his teachers. He struggled through trigonometry and his first semester of precalculus. Fortunately, his precalculus teacher provided extra support, working with him after school until he understood concepts. The teacher made Steve do problems on the board and explain his thinking. According to Steve, this teacher was the first mathematics teacher who "took interest in my success and encouraged me in a positive way." Steve eventually did well in his precalculus class, and the experience encouraged him to pursue mathematics-related courses in computer programming. Later, in college, Steve decided to major in cognitive science. He excelled in the calculus, biology, and statistics courses required for the major.

Although Steve acknowledged that his mathematics experience was "largely positive," he pointed to several issues that affected his experience and image of himself as a math learner:

> I have come to believe that the lack of diversity in my math education has limited my conceptual understanding and success in the field. During my geometry course, I began to assume I was not a good mathematician because I did not learn the concepts the way they were presented to me, leading to a lower self-image. A diverse student population and an encouraging teacher could have greatly enriched my experience. Students from wholly different backgrounds than me could have brought more ideas to the table because of their life experiences.

Steve's frustration with geometry, lack of diversity in mathematics classes, and need for a patient teacher were factors in his experience that he did not want to replicate in his future mathematics teaching. On the basis of his experiences, he was determined that in his own teaching, he would encourage students to draw on different mathematical ideas and strategies to solve problems, that he would listen to and respect ideas, and that he would promote students' participation to support mathematics learning:

> Encouraging all students to actively participate is necessary. If efforts are made to include all students in a class, proper respect is given to all ideas, participation is demanded and rewarded from all students, then the educational environment will benefit all involved.

Amy Collins: Middle school science and mathematics teacher

Amy is a seventh-grade science teacher in an urban middle school. She also teaches mathematics "support" courses (math intervention classes) for students scoring below "proficient" on the annual state standardized tests. Amy attended public schools in "suburbia across America." The backgrounds of her peers were like her own: "Caucasian, English speaking, and middle class." Her parents were very supportive of her in school and of her interests. Her father, a former teacher, excelled in mathematics and inspired her to follow in his footsteps. Amy was identified early as a "gifted" student. This allowed her to experience two different mathematics settings in elementary school. In her "regular" classroom, mathematics was taught through a "drill and practice" approach, with students given worksheets with "hundreds of practice problems" to do at home. Amy also studied her flashcards and "aced" her tests. In her "gifted" math class, the focus was on problem solving. Students learned mathematics that was connected with their interests. For example, Amy had an early interest in architecture. She met architects and learned how to design homes, developing floor plans by calculating the area, and so on.

Amy loved mathematics and considered herself a mathematician in elementary school. However, her middle school and high school experiences in mathematics quickly challenged the positive mathematics identity that she had developed in elementary school. Amy's gendered experiences with mathematics altered her mathematics learner identity. Being a mathematician and a girl was not well accepted socially in her school.

My test scores from my other school determined my placements in my classes, which, like many at that time, were separated by ability level. I got to take algebra in middle school and relished its challenges, but I started to notice that my fellow students weren't sharing my enthusiasm for the subject, at least not the girls. Other girls would demurely defer to the boys when the teacher asked a question, and if I raised my hand, they [female classmates] raised their eyebrows. Apparently, it just wasn't done…. [In] high school, girls could be smart, but only in language arts, art, and maybe social studies. I was torn again between the choice of making friends and fitting in and challenging myself to reach my potential.

When her family moved again, and she enrolled at her new school, Amy was placed in the advanced math and science courses. She was a junior taking senior-level courses. Because the students in these advanced courses self-selected for their mutual interests in mathematics and science, Amy reflected, "I finally found 'myself' again." Although initially steered toward engineering by her counselor, Amy decided to pursue her "love of nature and biology" at the university level by majoring in wildlife biology. Her math and science skills proved helpful in the "weeder" calculus courses in this male-dominated field.

Beginning in middle school and continuing through high school, Amy's experiences are examples of how gendered learning experiences can play out and influence the development of mathematics identity. Amy's gendered experiences in mathematics classes led to her first career as a wildlife biologist and later to a second career as a middle school science and mathematics teacher. These experiences had a direct impact on Amy's beliefs about the type of mathematics teaching and learning environment that she desired for her middle school students. For example, she wanted her mathematics classroom to reflect her "gifted" mathematics class experiences. All her students would be encouraged to be mathematicians and problem solvers. Furthermore, she voiced a strong commitment to empower all children, especially girls, to be passionate about mathematics and science. These views are directly connected to how Amy envisions her effectiveness as a mathematics teacher:

> I recognize that as a teacher, I will have to be careful not to teach the way I was taught in my regular classroom only but embrace the ideas about learning I experienced in my gifted class. This is the gift I can give to all my students, and especially my girls.

Kelly Ramirez: Fourth-grade elementary teacher

Kelly Ramirez is a fourth-grade teacher at an urban elementary school. She grew up in a multiracial, bilingual (English and Spanish), working-class household. She described her elementary mathematics experiences as enjoyable. She liked solving math problems and being able to read long decimal numbers aloud in class. Her parents were supportive. However, she recalled her father's evening kitchen-table drill sessions on multiplication facts as one of the only times that she felt mathematics anxiety.

Kelly's family moved a lot. She went to three different high schools. Kelly's transitions to new schools always meant questions about her academic competencies based on her "Hispanic last name" and her family's mobility patterns:

> They usually placed me into the remedial class before I ever arrived at the school. No matter the transcripts that preceded me, full of "highly capable"–this and "gifted"–that. A child who moved nearly every year? With a Hispanic last name? They, the administrators, made swift decisions. Only after meeting my mother (the teacher) and father (the member of the clergy who was working on his PhD), would they make the switch and register me in a new class. So very early on I learned that proving myself academically meant that I had to overcome my name. I couldn't just compete with the Joneses. I had to excel beyond them.

Kelly rejected mathematics in high school. She excelled in her humanities and drama courses and began to stoke her passion for social justice issues in the community. She felt algebra was "stupid" and not relevant to her life. However, part of her angst came from the fact that she had to borrow a graphing calculator while her middle-class peers pulled out their shiny new calculators to do math problems. She completely disengaged from geometry the following year, only going to class after being punished for skipping. The saving grace of her tenth-grade year was her Advanced Placement (AP) physics teacher. Impressed by her standardized test scores, he asked Kelly to take his AP physics course, where she studied mathematical concepts through physics applications. The class even went to an amusement park to ride roller coasters to prepare for the AP physics exam.

By the time Kelly transferred to her third high school, her focus was on social justice and the performing arts. At this point, mathematics "did not seem to fit." No one showed Kelly that mathematics was embedded in and relevant to her interests:

> I found joy in the self-expression of music and art and drama. Math just didn't seem to fit. But no one took the time to show me the everyday math: the fractions that

came so easily when I read music, the statistics proving my points about underserved populations, or even the strict meter and syllabic count of the Bard.

A need to work with mathematics resurfaced when Kelly became manager of a non-profit community organization that focused on supporting urban youth. Here she was in charge of budgets, statistics, and communication to funders who wanted quantifiable data.

For Kelly, both her pursuit of community activism on behalf of youth and justice and her later move to a new career as a teacher were fueled in part by mathematics learning experiences in her school and work contexts. Kelly's commitment to equity and social justice is a part of her mathematics teacher identity:

I realized that this is what kids need to know. They need to be inspired to take on the world, to fight for what is right, to express themselves in art and music … *and* they need to know the numbers. They need to be able to provide evidential, numerical proof, to balance budgets, and to fully understand the systems that they so desperately want to change. I knew I needed to make a greater impact, so I decided to become a teacher.

Socialization into or out of Mathematics: What the Autobiographies Show

The autobiographical stories of these six new teachers provide explicit examples of how the teachers themselves were socialized into or out of mathematics during their K–12 experiences and the impact of those socializing experiences on their mathematics teacher identities. Clearly, these past mathematics identities as learners directly link to their present mathematics teacher identities. Powerful experiences connected with race, class, gender, and culture shaped their view of mathematics teaching and what they want for their students' experiences. Leslie's struggle against the model minority myth prevented her from interacting with her white, middle-class peers in mathematically meaningful ways. Steve felt that the lack of cultural diversity in his environment had a negative impact on his mathematics learning experience because people from different backgrounds were absent from mathematical discussions. Amy received consistent messages from her peers suggesting that girls were not supposed to excel in mathematics—a message that contributed to her feelings of isolation in middle school and high school. Kelly's socioeconomic background was made apparent as her wealthier peers pulled out new graphing calculators while she had to borrow her calculator from the school.

Some of these teachers had powerful memories tied to specific math topics or domains that shaped their mathematics learning identity. Some were memories of challenges related to mathematical topics—for example, Michael's memory of trying and failing to find alternative methods for division with fractions to prove his teacher wrong, and Shawna's experience of taking precalculus three times. Leslie's, Steve's, and Kelly's dislike of geometry had major impacts on their confidence and sense of identity as mathematics learners. Furthermore, for some of these teachers, mathematics was a critical factor in choosing a college major. Although Amy and Steve felt at ease with the mathematics that they encountered in their majors, Shawna, Leslie, and Michael changed their majors to avoid having to take mathematics.

These teachers also noted the different ways in which mathematics was made relevant—or not—in their lives. Shawna's parents provided specific everyday activities such as baking and sports to explore measurement and statistics. Michael attributed some of his mathematics skills to his participation in sports (for example, street football helped him learn multiples of seven) and reading the sports page. Kelly's physics teacher made concepts of force and motion come alive through firsthand experience with roller coasters. Two of the new teachers pointed to "missed opportunities" to make mathematics relevant. Michael's experience with construction solidified his sense that mathematics is an integral part of everyday life. However, he had found it "cold and meaningless" when studying precalculus, and Kelly identified many ways to connect mathematics with aspects of her pursuit of music, theater, and social justice. Both teachers emphasized that making mathematics relevant to students was a key component of their math teaching vision. By telling their stories about mathematics learning, all these teachers were able to reflect on how their own experiences as learners shaped their visions for teaching and practice, thus contributing in powerful ways to their developing mathematics teacher identities.

Power and Status of School Mathematics and Its Impact on Mathematics Teacher Identity

The math autobiographies highlight the powerful role that mathematics learning played in the lives of the six elementary and middle school teachers. All the teachers understood the importance of mathematics and how they felt when they excelled or experienced difficulties in mathematics in school. For Amy, experiences with mathematics learning fueled her passion for nature as a wildlife biologist. By contrast, Shawna, Leslie, and Steve associated their inability to do mathematics with low self-confidence and achievement.

Society commonly views mathematics as a valued and high-status subject (Gutiérrez 2013a). Schools perpetuate this perspective through the gatekeeping structures associated with students' access to mathematics, such as tracking policies, reliance on standardized test scores, and specialized programs such as Advanced Placement. As we have learned from the six teachers' stories, unlike any other subject in school, mathematics has far-reaching consequences on children's academic identities and life transitions, including their economic and educational access, career decisions, and civic engagement or activism (Boaler 2002, 2008; Gutiérrez 2013a; Oakes 2005). Furthermore, what mathematics is has changed over time, making quantitative literacy a critical competency, essential for making sense of the data, information, and technology that are a part of our daily actions and decision making. This gives mathematics power and status that are important to understand as they relate to mathematics teacher identity.

The power and status of school mathematics often manifest themselves in decisions about what content gets taught, to which students, and by which teachers. As the narratives of the six new teachers show, what gets taught in the mathematics classroom shapes the mathematics identities of both students and teachers.

To gain a sense of the significant role that the status of mathematics plays in a math teacher identity, consider "algebra for all eighth grade" policies implemented in districts across the country in the early 2000s and outlined as one of two pathways in the common core standards high school appendix (National Governors Association Center for Best Practices and Council of Chief State School Officers 2010). Spielhagen (2011) describes the way in which a school district in the southeastern United States "detracked" the mathematics curriculum by mandating such a policy. The district implemented the policy incrementally through various stages in an effort not to overwhelm the system. Yet, the intent was to dismantle previously existing structures that blocked access to higher education for many students. Many teachers across the grade levels initially opposed the idea. The status of algebra and who had access to it became critical issues. These concerns filtered down to elementary teachers, who were feeling the pressure to "accelerate" the mathematics content for students.

According to Spielhagen, some elementary teachers claimed "that some students could not respond to such enriched experiences because they were 'not ready' for the work" (p. 37). In fact, the teachers often cited mathematical "readiness" but were unable to provide a definition on request. However, some of these teachers tied social class and family structure to readiness by citing a lack of parental support and "depressed home situations." From these teachers' perspectives, students with low socioeconomic status would be unable to gain access to or engage in the enriched mathematics experiences that the teachers believed the algebra mandate necessitated. In the words of a third-grade teacher, "Have you seen where they live? We are lucky that they get here at all" (p. 57).

So why was there such strong elementary teacher opposition to a policy intended to make mathematics accessible to more students? The reason was that the policy disrupted privately and publicly held notions about who should gain access to high-status mathematics, and the proposed change caused teachers to question themselves as effective teachers for particular students. No longer would algebra be reserved for an elite few. Making algebra accessible meant shifting teachers' sense of their work and their knowledge about students (Drake, Spillane, and Hufferd-Ackles 2011). The deficit views of poor children and their families became apparent in the face of the demands and mandates for increased access to high-status math content, such as algebra. This change had a significant impact on the elementary teachers' identities as effective mathematics teachers and their sense of how they could best help their students.

Spielhagen's study also found that teachers criticized the algebra policy because it "lowered the bar" for a course that had been reserved for students labeled as "highly capable" and "college bound." As one middle school teacher claimed, "My goal [for the students] is AP Calculus. When I teach algebra, that's where my students are heading" (p. 55). It was hard for this middle school algebra teacher to decouple the trajectory of the "highly capable" students and the domain of algebra (precursor to AP Calculus).

The policy challenged her view of the content (which she believed would no longer be rigorous), her idea of who should have access to the content (students whom she regarded as non–college bound), and thus her own identity as an effective algebra teacher. When the power and status of mathematics are challenged, the ripple effect on a teacher's mathematics teacher identity is clear.

Conclusion

Our goal in this chapter has been to highlight specific experiences and factors that shape mathematics teacher identity. The new teachers' autobiographies offer stories of supports and challenges that clearly have had a powerful impact on their own math learner identities—and, not surprisingly, a significant impact on their vision of what they want for their own students. Although some of these new teachers want to replicate what they experienced, others want to offer their future students a different, more positive experience. The importance of understanding the complex factors that contribute to a teacher's feelings, decisions, and actions related to teaching mathematics cannot be overstated. Experiences that teachers have in professional development or as a result of policy changes also have a significant impact on their mathematics teacher identities. Opening up access to high-status or more rigorous content can have a ripple effect on teachers' feelings of effectiveness and confidence about their likelihood of success with their students. Teaching new content with new students means a shift in thinking about practice and how to help students learn mathematics, thus seriously affecting teachers' understanding of their place in the world and their core beliefs. These experiences have an impact on our telling of our stories as math teachers—our evolving mathematics teacher identities. To promote reflection on mathematics teacher identity, we have provided a mathematics autobiography activity and a teacher identity activity at www.nctm.org/more4u. Readers can complete these activities individually or in a group setting.

DISCUSSION QUESTIONS

1. What is your mathematics learning autobiography? What aspects of your own history with learning mathematics do you think have an impact on your views about teaching mathematics? What kind of math identity do you want your students to develop in your classroom?

2. What roles did race, class, gender, culture, or language play in your mathematics learning story? How do those experiences connect with what you have observed in your own students and their developing math identities?

3. Reflect on the new teacher stories in this chapter. What elements of these stories resonate with your own mathematics learning experience? Do you find similarities or differences across the six stories that connect with the way that you teach mathematics?

4. Think about a time when you faced a change in the mathematics curriculum (for example, a shift in policy or a new textbook). Reflect on your feelings about that change. Were you excited? Disappointed? Frustrated? What was the impact of these changes on your view of yourself as an effective math teacher?

Chapter 4

Are the Standards Worthy of My Students?

In this chapter, we address *standards* and *standards-based practices*. These two terms are likely to generate a host of reactions depending on how the conditions in which you teach support, constrain, or resist the demands called for in mathematics standards documents and standards-based practices. Our orientation to standards and standards-based practices is that they are political entities (Martin 2019; Tate 2004). They are not neutral or objective (Apple 1992; Klein 2007; Kirst and Bird 1997; Romberg 1992; Schoenfeld 2004; Tate 1994, 1995, 2004). They often reflect the prevailing politics in the larger society. All teachers must negotiate these politics—making teaching a political activity (Gutiérrez 2013b). One of our goals for this chapter is to help teachers reflect, negotiate, and subsequently act in ways that are liberating and humanizing. We link our discussion of standards and standards-based practices to the justice-oriented and humanizing approach to mathematics teaching and learning that we framed in Chapter 1.

First, we discuss standards as *slogan systems* (Apple 1992; Tate 2004) designed to appeal to particular constituencies. As slogan systems, they take on a range of meanings that are invoked to generate these appeals. Next, we attempt to shift the discourse from what standards and standards-based practices mean to a consideration of whether they are meaningful. We offer some minimal expectations that we believe contribute to making standards and standards-based practices meaningful. Third, we propose another shift in discussions of standards and standard-based practices by moving from using standards to determine which students are worthy of mathematics-related opportunities to asking whether a given set of standards and standards-based practices are worthy of students. Fourth, we discuss why we believe that critical reflections on standards and standards-based practices are incomplete without attending to the political projects of white supremacy, antiblackness, nationalism, xenophobia, and racial capitalism. Racial capitalism is the process of deriving social and economic value from the racial identity of another person (Leong 2013; Robinson 2000). Historically, these political projects have appropriated mathematics education reforms, including standards-based reforms, to further their interests and work against justice-oriented and humanizing mathematics education for historically marginalized learners. In closing the chapter, we provide some thoughts and questions about how standards and standards-based practices might contribute to a more humane and liberatory mathematics education.

The Lure of Standards and Standards-Based Practices

As we shift from one crisis discourse to another in mathematics education (Martin 2019; Schoenfeld 2004; Washington et al. 2012), standards and standards-based practices seem to offer a pathway toward mitigating inequities in mathematics education. The logic underlying this line of thought suggests that if we can just do standards and standards-based practices better and more effectively, then students, teachers, and society will benefit. As noted by critical education scholar Michael Apple (1992), one reason for the lure and appeal of standards as the remedy for all that ails mathematics education rests on the fact that standards often represent *slogan systems*, with characteristics such as these:

- Standards must have a sufficient degree of vagueness so that powerful groups or individuals, or those who would otherwise disagree, can fit under the umbrella.

- Standards need to be specific enough to offer something to practitioners here and now.

- Standards need to have the ability to charm. The style must be such that it "grabs us."

Much of the authority that is associated with standards also stems from the belief that they are research based, hence free from ideology and bias. But, as we outline here, standards are not apolitical. They serve particular interests and reflect the preferences and biases of those who develop them. Moreover, if the underlying research is rooted in deficit perspectives, then those perspectives are likely to be reflected in the standards. As pointed out by mathematics educator William Tate (2004), the word *standards* also draws some of its appeal from the belief that whatever appears under its umbrella is comprehensive, authoritative, and powerful in reach. Here are some of the associations with standards that Tate (2004) noted:

- Standards as vision of ideal practice

- Standards as essential knowledge in a field

- Standards as descriptors of student performance

- Standards as guides to align system components

- Standards as measurable goals for student learning

- Standards as clear curricular goals for teachers

- Standards as guides in measurement and accountability systems

- Standards as descriptors of system inputs

- Standards as mechanisms to discuss opportunity to learn prerequisites and conditions for systemic success

In reading over this list, you may wish to reflect on when and where you have encountered these meanings and how they have been used to generate support from your or your colleagues. Do any of these meanings appeal to you? Which meanings have the greatest impact on your work?

From *What Do Standards Mean?* to *Are the Standards Meaningful and Empowering?*

Although we raise critical questions about standards and standards-based practices, we do not oppose standards and standards-based practices. Our support is for standards and standards-based practices that are meaningful and empowering. Our notion of *meaningful* is linked to a set of minimal expectations: (1) standards and standards-based practices should apply to the broadest range of students with varying levels of resources and supports, allowing students to build on their cultural and community knowledge, and should not privilege some students and backgrounds over others; (2) standards and standards-based practices should not be unilaterally imposed on students and teachers; there should be processes that allow students, teachers, and families to codevelop or vet the content, aims, and goals of standards and standards-based practices (Martin et al. 2019); and (3) standards and standards-based practices should not cater to the demands of high-stakes testing regimes that produce racial and social hierarchies enforced by punitive measures against teachers and students. Our view of empowerment aligns with that proposed by mathematics educator Paul Ernest (2002):

> Empowerment is the gaining of power in particular domains of activity by individuals or groups and the processes of giving power to them, or processes that foster and facilitate their taking of power. Thus a discussion of mathematical empowerment concerns the aims of teaching mathematics and the objectives of learning mathematics. It also concerns the role of mathematics in the life of the individual learner and its impact on their school and wider social life, both in the present and in the future. Empowerment through mathematics necessitates a consideration of the development of the identity of learners and their potentiation through the development of mathematical and related capacities. (p. 1)

Ernest outlines three areas of empowerment that we believe meaningful standards and standards-based practices should support.

- **Mathematical empowerment** concerns the gaining of power over the language, skills, and practices of using and applying mathematics. This is the gaining of power over a relatively narrow domain (e.g., that of school mathematics).

- **Social empowerment** through mathematics concerns the ability to use mathematics to better one's life chances in study and work and to participate more fully in society through critical mathematical citizenship. Thus, it involves the gaining of power over a broader social domain, including the worlds of work, life, and social affairs.

- **Epistemological (knowledge) empowerment** concerns the individual's growth of confidence not only in using mathematics, but also a personal sense of power over the creation and validation of knowledge. This is a personal form of empowerment: the development of personal identity so as to become a more personally empowered person with growth of confidence and potentially enhanced empowerment in both the mathematical and social senses (and for the mathematics teacher—enhanced professional empowerment). (pp. 1–2)

We encourage teachers to reflect on the minimal expectations outlined here and whether the standards and standards-based practices in their own contexts empower students mathematically, socially, and epistemologically.

Are They Worthy? Shifting the Discourse of Standards

We recognize that all teachers, students, and families must negotiate externally imposed standards and standards-based practices within their own teaching and learning contexts. These negotiations are often in response to the expectations of those who designed the standards and the preferred practices associated with those standards. For example, content standards presented by grade level assume and expect that students will enter particular grades having "mastered" particular skills. But are these expectations rooted in the belief that students come to school as brilliant and capable? Or do these expectations reflect beliefs that some students and their skills and competencies are in need of repair? Likewise, delineations of standards-based practices (e.g., Common Core Mathematical Practices; Process Standards—Texas Essential Knowledge and Skills for Mathematics; National Council of Teachers of Mathematics [NCTM] Principles to Action) assume and expect that teachers can interpret and actualize these practices with fidelity.

We encourage teachers to rethink their engagement with standards and standards-based practices. We offer some *different* considerations that we believe are consistent with the justice-oriented and humanizing approach to mathematics teaching and learning that we discussed in Chapter 1. We contrast the kinds of considerations and questions that we raise with those that are typically invoked when standards and standards-based practices are couched in equity-oriented rhetoric. For example, teachers are often asked to consider and act on questions such as the one presented below in service to efforts focused on equity and inclusion:

- How can we use standards and standards-based practices to help underrepresented learners gain the necessary skills and knowledge to take advantage of various educational opportunities, such as access to advanced math courses, STEM opportunities, and preparation for college and math-related careers?

Who could argue or take issue with standards and standards-based practices being framed in this way? This question appeals to teachers' senses of benevolence and their commitments to helping underrepresented students (e.g., Black, Latinx, Indigenous, poor) achieve upward mobility and social and economic assimilation. A deeper probing of this question, however, reveals how standards and standards-based practices are often used to determine *which learners are worthy* of these socially valued opportunities. Students who meet or exceed the standards measured by high-stakes assessments are frequently deemed more worthy than students who do not meet the standards or students who struggle to engage in standards-based practices. In essence, standards and standard-based practices become mechanisms of identity-making; teachers and students must then negotiate the identities that are made available and imposed on them. Teachers are deemed "good" teachers when a majority of their students meet the standards. Students are deemed "smart" or "struggling" or "below average" on the basis of whether they do or do not surpass the minimum standards.

We encourage teachers to stop discussing the *worthiness of students in relation to the standards* and standards-based practices and move to considering the *worthiness of standards in relation to students*. Here are some questions teachers can ask:

- Which standards and standards-based practices are worthy of the students who show up in my classroom?
- How comfortable am I when applying standards to children from backgrounds different from my own only to conclude that these children need repair?
- Are these standards and standards-based practices capable of capturing the brilliance of my students?

In our view, these questions invoke a do-no-harm stance toward students and teachers. Standards and standards-based practices that are truly worthy of students and teachers do not impose damaging identities on students and do not put teachers in positions to conclude that some children are broken and in need of repair. In our view, systems and structures that employ standards and standard-based practices in ways that require teachers to do so should be dismantled.

Standards and Standards-Based Practices as Political Entities

In reality, teachers are expected to further the interests of the prevailing systems and structures by implementing the standards and standards-based practices in their classrooms. Standards, for example, are frequently attached to opportunities for educational, social, and economic advancement that serve as incentives for students, parents, and caregivers. Standards can also reify hierarchies. For example, the Chicago Public School system (2023; 2024), in addition to having preschool testing for access to accelerated curriculum in kindergarten (access to what are called classical elementary schools and regional gifted centers), also employs an algebra exit exam, which they describe as follows:

> The Algebra Exit Exam is an annual assessment administered in the spring to all eligible middle-grade students enrolled in an algebra course. The results of the exam, along with other factors, are used to determine placement or credit at the

high school level.... The Algebra Exit Exam results are reported on the scale from about 150-415 and are based on the Illinois State Learning Standards and the CPS Algebra I course framework. The results are reported as Pass/Did Not Pass. A score of 300 and above is considered a Pass. A student with a Pass demonstrates a generally consistent command of algebra and can apply it in a variety of settings. (Chicago Public Schools 2023, p. 1)

The Algebra Exit Exam is one part of a more complicated and convoluted system that divides and sorts students among various types of high schools, with only a small percentage gaining access to those Chicago high schools considered by some to be among the best not only in the state of Illinois but in the entire country. The system of testing and its associated standards produces winners and losers in the competition for these schools.

Veteran teachers who have taught for more than twenty years will also recognize the current historical moment in mathematics education—which will also produce winners and losers, serve workforce and human capital needs, and reify social hierarchies—is not unlike other historical moments when standards received national attention and shaped teacher practice and the possibilities of teaching. Many teachers can recall that the NCTM produced the *Curriculum and Evaluation Standards for School Mathematics* (NCTM 1989), *Principles and Standards for School Mathematics* (NCTM 2000), and *Principles to Actions: Ensuring Mathematics Success for All* (NCTM 2014), three widely discussed and influential standards documents. In addition to laying out the content that was supposed to be learned in each grade and the practices in which students were expected to engage, each of these documents was characterized by an explicit focus on equity and inclusion as an acknowledgment of the status of marginalized learners in mathematics education. Each of these documents was also the subject of much debate among mathematics educators, mathematicians, policymakers, and parents. Debates about the purposes, goals, pedagogies, and practices of school mathematics, along with issues of access, became so heated that they were characterized as the "Math Wars" (Schoenfeld 2004).

Many readers are also aware that, in 2009, the Common Core State Standards Initiative was launched as an attempt to produce a set of standards that would be adopted by each state. With respect to inclusion and equity, these standards are undergirded by a "rising tide lifts all boats" ideology and philosophy, proposing that all students be held to high standards and that a concerted effort to achieve these standards will benefit all children. Like earlier standards movements, the Common Core standards have not been without controversy. More recently, as the realities of Common Core have set in (e.g., lack of support for teachers, developmentally inappropriate demands on children, alignment to state assessments, and parent frustration), some states have reverted back to developing their own standards. This retreat from Common Core is all too predictable for those who have studied the history of mathematics education reform in the United States and for teachers who have been around long enough to see one reform after another.

Along with various curriculum and content standards, we have seen a proliferation of standards-based practices, evidenced within the earlier NCTM documents but also in recent documents like *Principles to Actions*, published in 2014. The key practices from that document are meant to empower all students. The five equity-based practices discussed in this book have been used to support standards-based instruction for the new generation of standards-based resources including *Catalyzing Change in High School*

Mathematics (NCTM 2018) and "Catalyzing Change for Elementary School" (Huinker 2019). However, a close look at those resources suggests that the mere presence of equity-based practices is not the same as prioritizing.

This constant shifting (which precedes even the NCTM Standards) supports our claim that standards are political entities. As the political winds shift, so do standards.

Political Realities of Standards

One of the ideal visions for standards and standards-based practices proposes that when they are coupled with concerns for equity, inclusion, and diversity, and the benevolence of those who implement them, they can oppose those forces (e.g., white supremacy, antiblackness, nationalism, xenophobia, racial capitalism) that create dehumanizing and violent forms of mathematics education for marginalized learners (see figure 4.1).

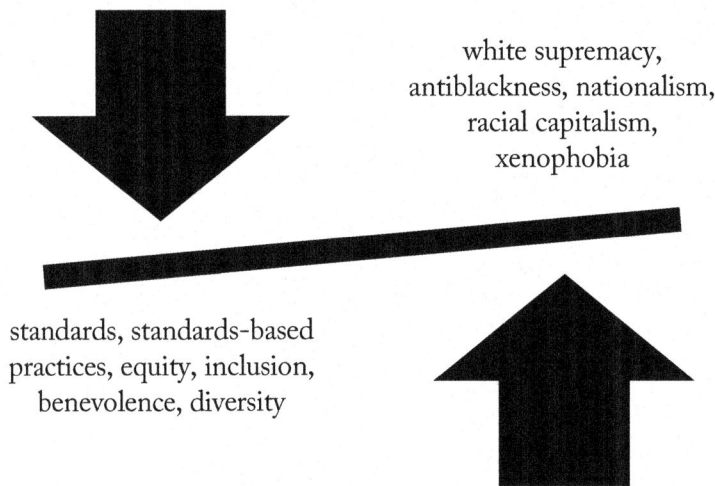

white supremacy, antiblackness, nationalism, racial capitalism, xenophobia

standards, standards-based practices, equity, inclusion, benevolence, diversity

Fig. 4.1. Standards and standards-based practices in opposition to oppression

We invoke two cautions about this ideal practice vision. First there is no evidence in mathematics education that standards, equity, inclusion, and benevolence are the antidote or counterbalance to white supremacy, antiblackness, nationalism, racial capitalism, and xenophobia (Martin 2019). As we reflect critically on the equity-oriented arguments in the first edition of this book, we realize that for some equity-minded individuals:

Equity work in mainstream mathematics education often represents little more than a convenient and comfortable waypoint so that the path of racial justice does not have to be traversed. Moreover, the [remedies and] forms of inclusion offered up in equity-oriented discourses and reforms have typically involved two trajectories: (1) inclusion accompanied by marginalization; and (2) assimilation into the existing cultures of mathematics education, thereby sustaining the fundamental character of the domain. (Martin 2019, pp. 460-461)

Mere inclusion or increased representation does not itself mitigate powerful forces of white supremacy, antiblackness, patriarchy, and xenophobia. Consider this thought experiment: Suppose that the commitment to standards, equity, inclusion, and benevolence was so powerful that every marginalized student exceeded the standards that regulate access to mathematical and other opportunities. Should students and parents who are marginalized in mathematics trust our existing systems to accommodate them? In this scenario, historical evidence shows that oppressive systems self-correct to mitigate these gains associated with inclusion, which can then result in maintaining the status quo (Anderson 2003; Bell 1980; Martin 2019). Consider the fact that women now make up over 40 percent of the math majors in university math departments. White women constitute the majority of these math majors awarded to women at 61 percent (National Center for Science and Engineering Statistics 2020). This inclusion has not extended in the same ways for Black, Latinx, Indigenous, and some women from pan-Asian backgrounds. And when they are included in mathematics-related spaces, women and girls in these groups often report hostile climates and attempts to further marginalize them (e.g., Borum and Walker 2012; Garcia, Rincón, and Hinojosa 2021; Joseph, Hailu, and Boston 2017). Research also shows that when attempts to expand curriculum opportunities and provide access to advanced coursework in elementary and high schools are made for Black and Latinx students, white middle-class and upper-class parents mobilize to protect these opportunities for their own children (Orfield 2012; Lewis, Diamond, and Forman 2015; Lyken-Segosebe and Hinz 2015).

We also ask teachers to think about a second-level caution about standards and standards-based practices. This caution focuses on the ways that standards and standards-based practices are put in service to and work to support oppressive political agendas outside of mathematics education (see figure 4.2).

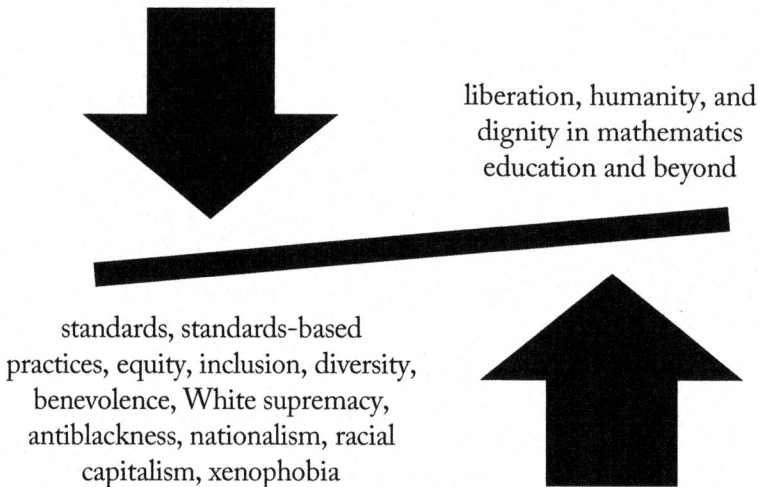

liberation, humanity, and dignity in mathematics education and beyond

standards, standards-based practices, equity, inclusion, diversity, benevolence, White supremacy, antiblackness, nationalism, racial capitalism, xenophobia

**Fig. 4.2. Standards and standards-based practices
in concert with oppressive forces**

We are aware that some readers may be concerned about or disagree with our alignment of standards with white supremacy, antiblackness, xenophobia, and racial capitalism. You might suggest that we are attributing negative intent to those who develop and implement standards and standards-based practices. We are less concerned with intent and more concerned with impact and the fact that standards and standards-based practices are often embedded in systems and institutions that work against liberatory and humanizing mathematics education. Our educational system is a structured caste system founded on institutional racism and wealth inequality (e.g., Darity 2022). The overall functioning of this system is not likely to be changed by doing standards and standards-based practices better. Indeed, even as one standards movement has shifted to another, the relative learning outcomes of Black, Latinx, and Indigenous students have not changed. And it is these groups of students that make up the majority of students in our public schools (National Center for Education Statistics 2022).

You might also be asking yourself: Is it fair or reasonable to place standards in opposition to liberating and humanizing mathematics education? We do not present figure 4.1 and figure 4.2 as the only two ways to think about standards and standards-based practices in relation to liberating and humanizing mathematics education. They serve as two important points of focus. We refer readers to Martin (2019) for a more detailed accounting of how standards and standards-based reforms have interacted with larger political considerations throughout history.

However, in thinking more deeply about figure 4.2, what is it about Black liberation and humanity, for example, that could provoke opposition so insidious that it draws on standards to do the work of oppression? Consider this framing of Black liberation that comes from writers William C. Anderson and Zoe Samudzi, in their book *As Black as Resistance: Finding the Conditions for Liberation.*

Black liberation poses an existential threat to white supremacy because the existence of free Black people necessitates a complete transformation and destruction of this settler state. The United States cannot exist without Black subjection, and, in this way, articulated racial formations revolve in large part around anti-Black regulations. (Anderson and Samudzi 2018, p. 15)

In this quote, the authors remind us that Black oppression, for example, is sutured into the systems, institutions, and regulations of the United States. The antiblack *regulations* they refer to typically utilize standards of some sort to regulate Black bodies, possibilities, and freedoms, particularly in school contexts. These punitive regulations have been applied to Black styles of hair, dress, and other embodiments of Black identities.

As we engaged in writing the new edition of this book, the nation was in the midst of social upheaval created by two social crises that have disproportionately affected Indigenous, Black, Latinx, and Asian communities: COVID-19 and racism. Deeply entrenched white supremacy and antiblack racism have been amplified by the extrajudicial killings of George Floyd, Ahmaud Arbery, Breonna Taylor, and countless others. Protests in numerous cities around the country helped to further highlight that the standards of justice for white and Black citizens, for example, are inherently unequal.

We extend this conversation with an example from mathematics education. But first, consider the following questions:

> - What do you think **should happen** when Black children exceed existing grade-level standards in mathematics?
> - What do you think **actually happens** when Black children exceed existing grade-level standards in mathematics?

In response to the first question, we hope that most teachers would respond by saying that Black students should be able to actualize all the opportunities that are associated with exceeding the standards. In response to the second question, we refer to a study published in 2014 by mathematics education researchers at North Carolina State University (Faulkner et al. 2014). This study was a longitudinal analysis of the different mathematics placement profiles of Black students and white students from late elementary school through eighth grade. This study utilized the Early Childhood Longitudinal Study—Kindergarten Class of 1998–1999 data set to analyze the impact of teacher evaluation of student performance versus student-demonstrated performance on the odds of being placed into algebra in the eighth grade. Data were drawn from third-grade math performance scores, fifth-grade math performance scores and teacher evaluation of student performance, and eighth-grade math course placement.

Results of the study revealed that Black students had reduced odds of being placed in algebra by the time they entered eighth grade even after controlling for performance in mathematics. Teacher evaluations of student performance were shown to play a greater and more adverse role for Black students than for their peers. So even if Black students performed above standards, teacher evaluations remained a key factor in preventing students to be appropriately placed in eighth-grade algebra. The odds of placement in algebra by the eighth grade for Black students were reduced by two-thirds to two-fifths compared with their white peers. Although the authors, by their own admission, did not enter the study with the intention of discussing antiblackness, they concluded:

> Black students confront an untenable impediment in that their Blackness (or, as we suggest here, the teachers' implicit responses to these students' Blackness) serves as an invisible, albeit formidable, obstacle to gaining access to higher level mathematics courses, irrespective of their demonstrated performance. (p. 306)

Exceeding standards, even when measured by accepted means of performance such as standardized test scores, did not serve as a counterbalance to antiblackness and white supremacy. In response to this example, we return to a version of the question that we posed earlier: What standards and standards-based practices are worthy of the children who show up in my classroom? We encourage teachers to grapple with this question knowing that all standards, no matter how well intentioned, are often embedded in systems and institutions that work against liberation and humanizing mathematics education. The overall functioning of these systems is not likely to be changed by doing standards better or even having more students meet the standards.

Toward Humanizing and Liberatory Standards and Standards-Based Practices

We reiterate that our goal in this chapter is not to critique any particular set of standards or standards-based practices. Our goal is to have teachers critically reflect on the question, Are the standards worthy of my students? We raise this question in service to more humanizing and liberatory forms of mathematics education. In seeking to address this question, we believe that the five equity-based practices framework described in detail in part 2 can help teachers critically analyze existing standards and standards-based practices.

- **Going deep with mathematics:** Do these standards and standards-based practices provide opportunities for teachers and students to engage in deep and meaningful exploration of conceptually rich mathematics content? Do these standards and standards-based practices provide contexts for high cognitive demand tasks and practices?

- **Leveraging multiple mathematical competencies:** Do these standards and standards-based practices honor and value multiple mathematical competencies and provide multiple points of entry into classroom practices?

- **Affirming mathematics learners' identities:** Can these standards and standards-based practices be implemented in ways that affirm a wide range of productive mathematical identities? Do they afford opportunities for all students to gain access to identities as doers of mathematics?

- **Challenging spaces of marginality:** Do these standards and standards-based practices assume skills and competencies that produce and naturalize race-, class-, and gender-based hierarchies?

- **Drawing on multiple resources of knowledge:** Do these standards and standards-based practices acknowledge and allow students to draw on their existing linguistic, cultural, and mathematical knowledge?

To this list, we add a sixth consideration that is consistent with our efforts to be more explicit about the larger sociopolitical contexts and considerations that facilitate or constrain teachers' practice and reflection (Gutiérrez 2013a, 2013b).

- **Sociopolitical context of standards:** What educational and political agendas are being served by these standards and standards-based practices? Whose interests are being served by these standards? Am I, as a teacher, being asked to teach and assess students in ways that dehumanize my students or view them in deficit-oriented ways?

Clearly, the good work that teachers of mathematics are doing in real classrooms every day must continue. Our belief is that there is a need to move beyond compliance with standards and standard-based practices to *commitments* and *obligations*. We offer some commitments and obligations that you can take up on your own paths toward a more humanizing and liberatory mathematics education:

- Respect and value the *humanity* of your students and their caregivers.

- Respect and value *childhood* for your students.

- Respect and value the *brilliance* of your students.

- Fulfill your obligation and commitment to do no harm to your students.

- Refuse standards and standards-based practices that attempt to determine which students are worthy.

- Commit yourself to standards and standards-based practices that are worthy of your students.

Conclusion

In this chapter, we encouraged you to rethink your engagement with standards and standards-based practices. Standards and standards-based practices are not neutral. They are invoked for political purposes, and they are linked to larger political agendas that are antithetical to liberatory and humanizing mathematics education. In our view, teachers can negotiate these politics by moving away from using standards and standards-based practices to determine which students are worthy of meaningful mathematics opportunities to asking whether standards and standards-based practices are worthy of their students.

DISCUSSION QUESTIONS

1. Which ideas in this chapter resonate most with you? Why? Which ideas cause you some level of concern? Why?

2. If you examine the standards and standards-based practices used in your local context, which of those standards and practices would you consider worthy of your students? Do they empower your students socially, mathematically, and epistemologically?

3. If you had to formulate your own standards-based practices that support humanizing and liberatory mathematics education, what would they be? Why do you believe these practices are humanizing and liberatory? How would you explain the value and relevance of these standards to students? To colleagues? To parents and caregivers?

Chapter 5

Math Strong: Reframing Beliefs and Structures to Disrupt Deficit-Based Thinking

As you learned in Chapter 2, mathematics identity refers to "the dispositions and deeply held beliefs that students develop about their ability to participate and perform effectively in mathematical contexts and to use mathematics in powerful ways across the context of their lives." Math identities are dynamic, constantly being negotiated, and include self-understandings as well as how others position you as a math learner and doer. Math identity is also just one of many identities we develop and continue to negotiate. Yet, it is one of the most powerful identities we have given its connection to how we currently measure academic "readiness" and "success" with high-stakes standards tests and through oppressive structures of schooling such as tracking that rank, sort, and segregate students. For example, Boaler and Selling (2017) published the results of a longitudinal study conducted with participants who, in a previous study, experienced two very different approaches to high school mathematics instruction. One high school taught mathematics in a tightly controlled, tracked setting with fixed-ability groups. The other high school was detracked and regularly supported collaborative learning in mixed-ability groups. Now young adults, students expressed the far-reaching impacts of those two types of experiences on their livelihood and well-being. The tracked setting was described as "psychological imprisonment," where students felt limited agency and authority over the mathematics they studied. The mathematics emphasized procedural fluency with little connection to real-life applications. The detracked setting offered "intellectual freedom" with many benefits, such as an emphasis on problem solving, questioning, and creativity that were useful beyond the classroom setting. The results further demonstrate that mathematical experiences in school settings have far-reaching impacts on how people feel about themselves as doers of mathematics and the role mathematics may play in their employment, education, and life choices.

The results of Boaler and Selling's study raise the question about what teachers are doing to construct experiences that students describe as fostering intellectual freedom or violence. In this chapter, we will introduce an identity concept called *Math Strong* (Aguirre 2016). We will describe a set of activities that helps us think critically about our role as identity workers in shaping mathematics identities of children. Our goal is to help you interrogate beliefs and structures that support or inhibit positive mathematics identity development in children—a Math Strong identity. We will conclude the chapter by making connections to the five equity-based practices.

Activity 1: Math Strong Brainstorm

There are many identity-linked phrases that connect to a specific experience or set of experiences. For example, have you heard of *Live Strong*? Live Strong is a nonprofit foundation focused on supporting people living with cancer. Have you heard of *Boston Strong*? Boston Strong is a phrase that originated after the horrific bombing at the Boston Marathon on April 15, 2013. In both cases, the phrases generate thoughts of unity, strength, support, and solidarity, often in the face of adversity.

Consider the phrase *Math Strong*. What kinds of words or descriptions come to mind? Take a moment to write down as many words as you can that you associate with Math Strong. In the context of thinking about teaching young people, here is a sentence starter:

"Math Strong students are …"

Figure 5.1 shows a word cloud representation generated in professional learning community when asked to describe Math Strong students. The size of the word reflects how many times the word was used as a descriptor. These words paint a picture of Math Strong students as being creative thinkers, curious, collaborative, and persistent. Positive

Fig. 5.1. Math Strong Word cloud

descriptors include resilient, flexible, and brave. This activity is designed to surface beliefs about mathematics learning that are aspirational. Teachers could ask themselves how their instruction supports these positive characteristics.

As teachers of mathematics, we are all identity workers. Rochelle Gutiérrez (2013b) describes this as follows:

> By virtue of mathematics being political, all mathematics teaching is political. All mathematics teachers are identity workers whether they consider themselves as such or not. They contribute to the identities students construct as well as constantly reproduce what mathematics is and how people might relate to it (or not). (p. 11)

As identity workers, our practices within systems contribute to how we teach mathematics and how students experience mathematics. This means that we have a professional responsibility to look carefully at our practices and the systems in which we teach and make changes—creating spaces for innovation, collaboration, and resilience. This is the essence of being Math Strong. However, nurturing Math Strong students is complicated by the unjust and inequitable systems in which we teach. As highlighted in previous chapters, standards, tracking, and testing can affect the identity work of teachers. We suggest here that all teachers should ask the following question: How can I cultivate Math Strong students in an unjust, inequitable system of mathematics education?

We believe that to adopt a Math Strong teaching orientation, especially in relation to students who are not assigned productive and positive math identities, one must interrogate the beliefs and structures that shape mathematics teaching and learning in the classroom. In our professional development work with teachers, we have found that the activity described next has proven to be helpful in moving practice toward a more humanizing and just mathematics education (for more resources on humanizing mathematics education, see Goffney and Gutiérrez [2018]). It is designed to surface beliefs about the children we teach. We started this activity in 2015 with a group of educators in Chicago. Since then, we've done it in many large and small professional development venues.

Activity 2: What Have You Heard? A Beliefs-Surfacing Activity

This activity is also designed to be done in professional learning communities.

Step 1: Create individual posters with descriptors like those shown in figure 5.2. The descriptors you select should reflect the different group designations of children you work with. These descriptors can be obtained from official documents in your school context. It is important to include variety. The descriptors should relate to class, gender, race, ethnicity, immigration status, religion, language, and region.

Step 2: Answer the essential question: What have you heard about working with these groups of students? Each person should be given sticky notes. For each poster, participants should write down one association per note and stick it on the corresponding poster. There is no limit on the number of sticky notes for each poster. Give participants enough time to generate a number of notes per poster (approximately 10–15 minutes). It is important to state in the instructions that people can write down what they have heard. It is not necessary to believe in or agree with the statement they write down. It is also OK if people have not heard much about some of the particular

What have you heard about working with these
groups of students in math?

- English language
  learners
- Black boys
- Asian/Asian
  American children
- Poor children
- Latinx children
- White children
- Military children

- Native American/
  Indigenous children
- Children with
  disabilities
- Vietnamese children
- Rural students
- Immigrant children
- LGBTQ+ Children
- Russian children

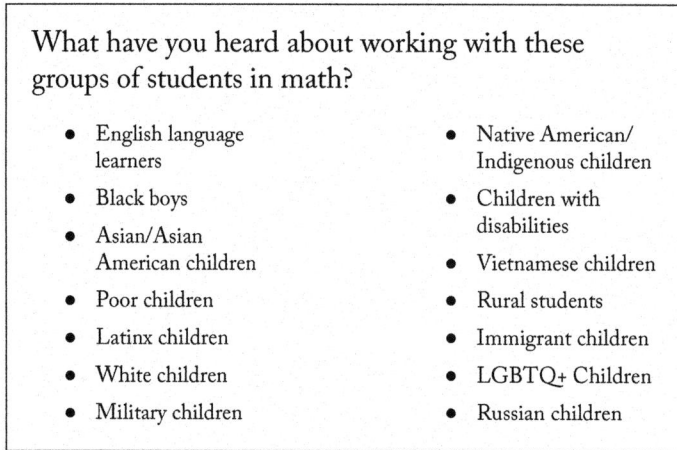

**Fig. 5.2. Sample student group names for
"What Have You Heard?" activity**

groups. Both the associations written on the notes and the quantity of notes on each
poster can be sources of inquiry in the next part of the activity.

Step 3: Look for patterns and themes. In small groups, explore one or two posters. For
each poster, identify and summarize important themes you notice. The notes allow you
to group ideas in cluster themes. As you identify your themes, think about the following
questions:

- If you were to put these themes in big buckets, like positive and negative,
  how might you reflect that information in percentages? Would it be
  50 percent positive, 50 percent negative? Are the data mostly positive?
  Mostly negative?

- What does the quantity of notes mean in relation to mathematics teaching
  and learning of that specific group of children? What if there are very few
  statements made about a particular group? What does that say about the vis-
  ibility of the student group in relation to mathematics teaching and learning?

- If you were to use the air quality index colors to evaluate the poster (e.g.,
  green is healthy, yellow is moderate, orange is unhealthy for sensitive groups,
  red is unhealthy, purple is very unhealthy, and maroon is toxic), what color
  would you assign the poster?

This part of the activity concludes with groups reporting their findings to the whole
group. Then, the conversation can be expanded to explore broader themes related to
questions such as:

- What similarities or differences do you notice across groups?

- Were there belief statements that appeared on several posters?

- What if you were a recent immigrant Vietnamese girl from a low-income
  household? How would you be positioned?

- How is intersectionality reflected in these posters, given that many of our
  students are a member of more than one of these subgroups?

A closer look

In our experience with this activity, deficit thinking dominates the statements on these posters. A closer look often reveals racist, sexist, classist, and other harmful stereotypes about the children we teach. Table 5.1 summarizes data from the "What Have You Heard?" activity conducted at the 2016 TODOS: Mathematics for All conference in Phoenix, Arizona. The student groups focused on here are poor students, immigrant students, Asian students, Latinx students, American Indian students, and Black boys. For clarity, n is the number of notes collected on the poster for each group of students. There are representative statements that were on the posters to reflect some themes. As you read through these examples, take time to compare these statements with what is written on your own posters. Are there similar statements? Similar percentages of positive and negative comments?

Table 5.1. "What Have You Heard?" activity statement data for selected student groups

Student group	n	Positive (+)	Negative (−)	Neutral (0)	Example statements
Poor	82	6%	89%	5%	(+) They *can* succeed in school; Need opportunities; Parents work long hours, multiple jobs, still care greatly about child's education; Redefine what supports mean by affirming their values
					(−) They don't need advanced math; They need work skills, not college skills; They have no experiences; Weak background knowledge; They have so much to worry about, can't learn; Parents don't care; Parents uninvolved; Parents can't/won't help with homework; Too late, why waste your time
					(0) Free lunch, breakfast in classroom, summer meals
Immigrant	64	15%	81%	4%	(+) Multilingual; Culturally specific algorithms; Hard working; Excels in math; Loves science
					(−) They don't speak English well, so learning math is an issue; Language is a barrier; A student is labeled at-risk in my district if the student is bilingual; All immigrant kids are uneducated; You have to dumb things down because they don't have the background; No or low parent involvement; parents uneducated; Very little mathematical knowledge
					(0) Math is universal; All have the same learning profile or background

(continued)

Table 5.1. "What Have You Heard?" activity statement data for selected student groups (*Continued*)

Student group	*n*	Positive (+)	Negative (–)	Neutral (0)	Example statements
Asian	68	94%	6%	0%	(+) Smart in math; Good in math; Smart/gifted; Innovative; Model minority; Respectful; Driven; Hard workers; Parents care, strict, value education (–) Work so hard, but have zero creativity; Not creative; Won't look you in the eye; Overscheduled
Latinx	54	18%	78%	4%	(+) They know more math because they already learned it in Mexico; Hardworking; Appreciative families (–) They can't learn math if they don't speak the language; Unruly; Gang members; Lazy; Lack motivation; Their families and culture don't value education; Latina girls don't think about college because they are going to be moms (0) Transnational; Latino students are English language learners
American Indian	45	7%	93%	0%	(+) Capable; Respectful (–) Native youth are not good at math; Native people do not think mathematically; Slow/stupid; Lazy kids, Lazy parents; Families on "the Rez" don't value education
Black boys	59	7%	93%	0%	(+) They tend to like to find their own way to solve a problem; Black boys have much potential that needs to be cultivated! (–) Don't have the basic skills to learn high school math; Can't learn math; Athletic but not academic; Hostile/aggressive/contentious; Loud; Lazy; Family doesn't care about education; Their culture prevents them from learning

Poor students

Nationally, 52 percent of students enrolled in public schools are eligible for free and reduced lunch (National Center for Education Statistics 2021), with some states and municipalities exceeding 70 percent, such as District of Columbia (76.4 percent),

Mississippi (74.8 percent), and New Mexico (71.9 percent). As you can see from table 5.1, most of the association statements about poor children are negative. There were several comments that positioned poor students as not needing to take advanced mathematics, which might lead to college. Instead, "work skills" should be emphasized. There also seems to be a time limit to learning expressed in these statements, such as "too late, don't waste your time." Although it is unclear what "background knowledge" means, poor students are limited. And, in an extreme case, they are not worthy of educating. In addition, almost half of the negative statements focused on deficit views about family. Comments such as "parents don't care" or "can't help," especially with homework, were pervasive. However, there were a few positive statements made that provide a counter-narrative to the dominant deficit view. For example, "They *can* succeed in school" and "need opportunities." In relation to families, a few statements acknowledge that parents do care about their child's education while navigating difficult life circumstances that may limit their capacity to participate in traditional school-sanctioned parent activities. The last positive statement suggests that perhaps schools need to redefine what is meant by "supports" to better align with students' and, by extension, their families' values rather than the other way around.

Immigrant students

By U.S. law, all children, in light of their immigration status, are entitled to a free public education (*Plyler v Doe* 1982). According to U.S. Census data, approximately 25 percent (almost one in four) of children come from immigrant households (where at least one parent was born outside of the United States). Of this group, only 12 percent of the children were born outside of the United States (Migration Policy Institute 2018). The statements related to immigrant students also skew sharply negative. Many of the negative statements highlight language as a barrier to learning. More specifically, statements claim that not understanding English is a barrier to learning mathematics. One comment also reveals the systemic nature of deficit thinking by district policies and practices that label bilingual children as "at-risk." Deficit views about parents are also prevalent. Again, parent involvement and parent education are framed as seriously limited or nonexistent (e.g., "uneducated"). Another negative theme relates to knowledge. Immigrant children are framed as lacking background or mathematical knowledge to support their learning. On the other hand, the small number of positive comments suggests cultural and linguistic assets and enthusiasm for math and science learning.

Asian students

This term is broadly used in most school contexts to refer to students with ethnic backgrounds from countries of origin along the Pacific Rim as well as India. According to the National Center for Education Statistics (NCES) (2022), Asian students make up 5 percent of the public school enrollment. Although the overwhelming descriptors may seem positive (e.g., smart, and smart in math), these descriptors are grounded in racist stereotypes that do not recognize the diversity among Asian American ethnic groups (e.g., Korean, Chinese, Vietnamese, Filipino), intersectionalities (you can be poor, immigrant, and Asian), or individual agency. The model minority myth is incessant and has an enormous impact on children's academic and sociocultural identities (Lee 2009;

Shah 2019). Being labeled as smart or smart at math represented two-thirds of the positive statements collected. However, in this context, these types of comments are not a compliment but a racialized insult marking Asian youth because of their race. These stereotypes hurt as reflected in Baye's story in Chapter 1 and the elementary teacher, Leslie Park, in Chapter 3. The other positive comments focus on student behaviors and reflect an ethic of hard work and respect. Parents are also framed in a positive light as valuing education. With the vast majority of the statements making racialized claims of "smartness" as it relates to a particular racial/ethnic group, it is important to ask, What if you struggle with mathematics and you are a member of this group? How might you be dismissed? Ignored? Dehumanized?

Latinx students

Latinx students make up the fastest growing population of students enrolled in public schools. Twenty-eight percent of students enrolled in public schools are Latinx (NCES 2022). With respect to Latinx students, negative statements were pervasive. Like immigrant students, language is depicted as a barrier to learning mathematics. In addition, the behavior themes around laziness, motivation, and violence were called out. There were also a few gendered comments suggesting that Latinx girls choose parenthood over higher education. Comments about families were split in this group. There were a set of comments that families and culture do not value education or that family was important and school was not. While at the same time, some of the positive comments positioned families in more affirming ways: families are "appreciative"; they "sacrifice a lot" for their children's education. The few positive comments also identified positive behaviors such as "hardworking" and identified the benefits of previous schooling in other countries like Mexico that situates these students as knowing more mathematics through their country of origin.

Indigenous/American Indian students

The student enrollment data for American Indian/Native Alaskan is 1 percent of the public school population (NCES 2022). For Indigenous/American Indian students, the comments were overwhelmingly negative. One theme related to student capacity to learn mathematics with comments such as American Indian youth are "not good at math," "do not think mathematically," or "do not have good number sense." However, intelligence was questioned overall, with several comments stating that American Indian children were "slow/stupid," "slow learners," or "slowed down the class." Families were targeted with comments such as "lazy" or "don't value education." And there were comments related to the role of alcohol in families as a barrier to learning (e.g., "parents are just drunks anyway"). There were two positive comments that were affirming: Indigenous youth are "respectful" and "capable." The domination of deficit responses fuels racist stereotypes that pervade the educational landscape for Indigenous/American Indian youth and families.

Black boys

Black students make up 15 percent of the public school enrollment population (NCES 2022). Black children are disproportionately disciplined more than any group of students in public schools starting with preschool (Fabes et al. 2021). The persistence of negative deficit-based beliefs continues with the overwhelmingly negative comments about Black boys. Many of the comments focus on racist stereotypes that dehumanize and position Black boys as a threat, unintelligent, or unworthy. Some representative comments reiterate that family and culture are barriers to learning. There is a strong sense that Black boys cannot learn or advance in mathematics.

The above statements about Black boys (Table 5.1) were recorded in 2016 from an activity conducted in a large conference audience of math educators, including teachers, instructional coaches, and researchers. Table 5.2 shows association statements about Black boys collected from elementary and secondary preservice teachers enrolled in a university mathematics methods class in 2019. It is important to realize that these preservice teachers have just entered the profession through their field placements. They are hearing statements from practicing teachers and administrators.

Table 5.2. Preservice Teachers' Association statements about Black boys

Pipeline to prison	Not math people, not interested in math
Adult-ified	Poor math students
Live in the "ghetto"	Struggle with math
Low expectations	Candidates for high dropout rates
Don't feel they need math, want to be successful at sports	Held back
Lots of energy	Slow learners
Low attention	Not motivated/not ambitious
Families don't support	Don't care
Disruptive/aggressive	Haven't heard positive or negative things

We invite you to read this list aloud. How does this make you feel? Now, imagine if you were a Black boy entering class and these are the belief statements about you. These statements inflict violence on the well-being of this Black boy. What kind of math identity will these statements grow? How will this child become *Math Strong*?

What does this mean?

You may have noticed that across the posters, the depth and consistency of deficit thinking is remarkable. How could this be? How could we come up with such consistently negative, harsh, hurtful, mean-spirited descriptions of children from different racial, ethnic, class, gender, and language backgrounds?

When we conducted this activity with teachers, coaches, administrators, and teacher educators, some expressed extreme discomfort with the results. We have been told: "You tricked us; you set us up," and we have had participants walk out of the room and not return. The activity is designed to surface beliefs and promote critical dialogue about those beliefs that are in the air to be heard by others, including the children themselves. However, it is important to note that the "What Have You Heard?" activity was preceded by the Math Strong brainstorm activity. In addition, the instructions of the "What Have You Heard?" activity invited written statements that you could agree with or not. This means that as a participant, one has individual agency to write down positive affirming statements on the posters as well. What troubles people are the results. They are depressing, upsetting, and frustrating. Feelings of anger and shame may also be present.

We believe teachers did not become teachers to harm, hurt, or give up on children. It is important to know that the thickness of this deficit thinking is deeply integrated into our educational system and has been supported by educational research, especially in mathematics education. The connections are historical, cultural, institutional, and ideological. And when that research is reported to the press, it spreads to the public discourse. We have, in fact, normalized this deficit thinking as the default. Thus, we all breathe this smog of deficit thinking, and we have been doing so for decades.

The next section will provide a brief overview of why deficit thinking is so pervasive. It will highlight the role of white supremacy in school systems and educational research. This primer is to provide context for the prevalence of deficit thinking in mathematics education, while at the same time, we seem to be able to clearly articulate what we want to see in Math Strong learners. Then we will finish this chapter with a discussion on how to reframe these deficit views and take action that will lead to rethinking equity-based practices.

The Depth of Deficit Thinking

The construct of deficit thinking about those who have been historically denied access to basic education, which includes mathematics literacy, has historical, cultural, institutional, and ideological facets (Valencia 2010). Table 5.3 presents a timeline of deficit-based thinking in education.

Table 5.3. Timeline of deficit-based thinking in education

Structural oppression (nineteenth century)	Grounded in legal and institutional systems of slavery and segregation (e.g., compulsory ignorance laws in southern states between 1740 and 1832)
Psychometrics and educability (early twentieth century)	Strong connection between intelligence tests (e.g., Stanford-Binet) and the eugenics movement that institutionalized racial hierarchies with "measured" intelligence and systematized segregation of children into different academic tracks (early twentieth century; see Terman [1916])
Victim blaming (1970s)	Person-centered linked to group membership (e.g., racial minority and socioeconomic status); focused on alleged individual, cognitive, and motivational deficits, not on structural or systemic inequities
Gap gazing (late 1990s to present)	Hyperfocus on achievement and interventions based on standardized testing outcomes that label readiness, determine course placement, and structure educational trajectories.

Structural oppression of deficit-based thinking started prior to the Civil War in which laws were passed that made it illegal to teach an enslaved person to read. Literacy education for Black people was a threat to the system of oppression and therefore banned by law (Valencia 2010). In the early twentieth century, scholars used "scientific" approaches to justify eugenic claims of racial superiority of white northern Europeans, leading to the institutionalization of racial/ethnic hierarchies and systematized segregation of children into academic tracks and separate school systems (Berry, Ellis, and Hughes 2014; Ellis 2008). In the 1970s, deficit-based thinking was legitimized in research studies showing students from particular income levels and ethnic/cultural groups had cognitive and motivational deficits, blaming the individual children and their families for persistent underachievement (Valencia 2010). Since the late 1990s and with the ushering in of the No Child Left Behind Act, "gap gazing," or the hyperfocus on test scores in conjunction with demographics, has continued to reinforce structures and practices that label readiness, perpetuate ability grouping and curricular tracking practices, and reinforce educational trajectories with consequential and differential outcomes to income, career choices, and postsecondary education (Flores 2007; Gutiérrez 2008). As scholar Cacey Wells (2018) describes:

> Tracking students based on ability fuels *academic apartheid* in mathematics education, as tracking often includes reproduction of social class by creating modern systems of segregation. (p. 72, italics added)

Deficit thinking has been embedded in our educational system from the beginning. However, it is important to understand the role mathematics education research plays in the maintenance of deficit thinking about children of color and indigeneity and those affected by poverty. To illustrate this point, we will focus on two influential documents that summarize the research on early childhood mathematics. Below is a quote from an influential report on mathematics education called *Adding It Up* (National Research Council 2001a). This report defined the five strands of mathematical proficiency that were discussed in Chapter 2 of this book. However, this quote comes from another part of that report focusing on the mathematical knowledge children bring to school.

> Overall, the research shows that poor and minority children entering school *do possess some informal mathematical abilities* but that many of these abilities have developed at a *slower rate* than middle class children. This *immaturity* of their mathematical development *may account for the problems* that poor and minority children have in understanding the basis for simple arithmetic and simple word problems. (p. 73, italics added)

This quote positions "poor and minority children" as developmentally slower than wealthy and white children. Although they have some mathematical abilities, their abilities are "immature" and contribute to problems in their understanding of even "simple" mathematics. This statement reinforces an educational caste system based on structural racism and income inequality. It sets up math teachers to believe that children who are not white and wealthy will be slower to learn and harder to teach when they enter school. These research-based statements echo the association statements collected from the "What Have You Heard?" activity.

Other scholars have made similar claims in influential reports, such as the *Second Handbook of Research on Mathematics Teaching and Learning* (Lester 2007). Here is an excerpt from a chapter on early childhood mathematics (Clements and Sarama 2007):

> Into kindergarten and the primary grades, lower-income children more than middle-class children, *use less adaptive and even maladaptive strategies, probably revealing a deficit* in the intuitive knowledge of numbers and different strategies.... In summary, *although there is little direct evidence on this*, we believe a pattern of results suggests that, although low-income children have pre-mathematical knowledge, *they do lack* important components of mathematical knowledge. *They lack the ability—because they have been provided less support to learn—*to connect their informal pre-mathematical knowledge to school mathematics. (p. 534, italics added)

This quote focuses on social class, which can also be a proxy for race since Black, Latinx, and Indigenous students are overrepresented in communities affected by poverty (Milner 2013). Again, children from low-income and working-class backgrounds are positioned in relation to having cognitive deficits. Not only do these deficits include slower-developed forms of mathematical knowledge, but the use of the term *maladaptive strategies* suggests a pathology in the children's mathematical thinking and development. The last sentence makes note of premathematical knowledge being evident but insufficient to learn school mathematics because poor children have been "provided less support to learn." If we are talking about early childhood, where does premathematical knowledge come from? Although not explicitly stated, the inference is clear: the child's home. This quote places responsibility on families as the source of the mathematical knowledge problem perpetuating oppressive stereotypes such as parents don't care or are ill equipped to support their child's learning.

The third example comes from a 2015 article that got a lot of press coverage. The article focused on the cognitive development of Mexican American toddlers (twenty-four months old) and mothering practices (Fuller et al. 2015). Summarizing related research, the article states:

> Recent work details the *lagging cognitive functioning and linguistic proficiencies* of Latino children as they enter kindergarten. This includes Mexican American children who enter kindergarten with *weaker* pre-literacy skills (in English or Spanish), familiarity with print materials, and *knowledge of mathematical concepts*, relative to their White peers. (pp. 141–142, italics added)

In this quote, deficit-based views of children's cognition and language development as they enter school *centers on whiteness*. Latino children, and Mexican American children specifically, are found to have "weaker" knowledge of math concepts upon entering kindergarten. What is important to point out is that ethnic backgrounds are now being compared with racial backgrounds: Mexican American and white. There is no mention of the ethnic backgrounds of the "white peers" such as French American, Swedish American, or German American. In addition, the maternal practices of white mothers are elevated in this study. By disregarding ethnicity for white participants, whiteness is privileged and normalized echoing the eugenics frame of the early twentieth century in the findings of this study.

However, the most destructive aspect of this work is how it gets disseminated in the public media and enters the public discourse. For example, the headline for the National Public Radio article was "Mexican-American Toddlers: Understanding the Achievement Gap" (Sanchez 2015), and the headline in the *Los Angeles Times* was "Literacy Gap between Latino and White Toddlers Starts Early, Study Shows" (Watanabe 2015). These headlines amplify the familiar deficit refrain about Mexican American children while centering whiteness by using "achievement gap" language (Flores 2007; Gutiérrez 2008; Martin 2009a). Now, the education of white toddlers and the child-rearing practices of white mothers are elevated. The *Los Angeles Times* (Watanabe 2015) newspaper quote from the author of the study showcases the overt racist stereotypes embedded in this study: something is wrong with Mexican American children and their upbringing.

> "For many Latinos, the home is a nurturing and supportive environment, but it's not necessarily infused with rich language and cognitive challenges," Fuller said. "Being warm and fuzzy may lead to well-behaved youngsters, but it doesn't necessarily advance a young child's cognitive agility."

Studies such as the ones discussed here reinforce deficit views of children of color and those children from working-class/low-income backgrounds. Parental practices and children's cognition (key indicators of mathematical learning) are villainized. This recent scholarship reinforces the victim-blaming perspective and weaponizes the gap-gazing perspective that has led to the explosion of systemic "interventions" that children of color and those living with poverty have experienced in schools. Findings of these studies suggest that these children and mothers need to be fixed with the goal of being more like white middle-class parents and students for school and mathematics education success. The normalizing of whiteness and deficit framing of low-income children and children of color starts young and persists through K–12 schooling. This is the smog we breathe—everyone. And it shapes how we see ourselves and our children as learners inside and outside the classroom.

You have now experienced an activity that surfaces beliefs that are overwhelmingly negative about children and families we work with, rooted in persistent racist and classist stereotypes and white supremacy, supported by mathematics education research, and reinforced by structures such as high-stakes assessment systems and curricular tracking. The depth and breadth of deficit thinking surrounds us. And yet, you also have a clear vision of what you want in Math Strong students. How can we reconcile these two realities? We don't. We must disrupt current policies and practices that reinforce deficit thinking, segregation, and harm and replace them with equity-based policies and practices described in this book. It is now time for some reckoning and innovation to get us unstuck and moving forward in a positive direction toward liberation.

Cleaning the Air: Reframing toward Math Strong

One of the first things one can do is to participate in a reframing activity that begins to reorient our beliefs and connect to equity-based actions. In Chapter 2, we started this process by calling out common deficit statements and reframing them with strength-based language. Some examples were from "at-risk" to "resilient" and from "limited English speaker" to "multilingual language broker." The reframing language elevates the assets and agency students bring with them that can be nurtured in the classroom.

Figure 5.3 shows how this can be done using some of the deficit language present in the preservice poster for Black boys. For every deficit comment, a strength-based comment is created. Read the bolded list aloud. How do you feel? These beliefs statements reflect a completely different view of Black boys, one that amplifies positivity, strength, intellect, knowledge, family support, and leadership. Imagine if this was the list teachers had in their mind when a Black boy entered their classroom. Would these statements sustain a positive math identity? Would these statements foster a Math Strong student?

Reframe Activity:
Rewrite deficit statements using strength-based language

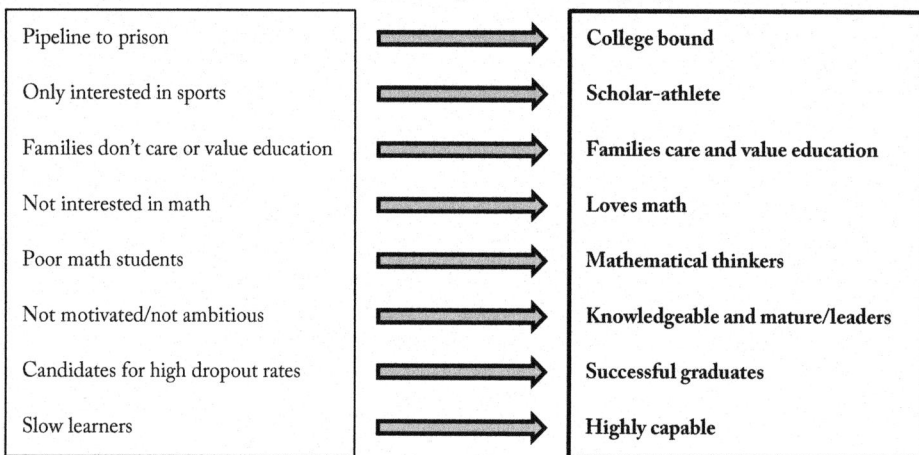

Deficit		Strength-based
Pipeline to prison	→	**College bound**
Only interested in sports	→	**Scholar-athlete**
Families don't care or value education	→	**Families care and value education**
Not interested in math	→	**Loves math**
Poor math students	→	**Mathematical thinkers**
Not motivated/not ambitious	→	**Knowledgeable and mature/leaders**
Candidates for high dropout rates	→	**Successful graduates**
Slow learners	→	**Highly capable**

Fig. 5.3. Reframing activity

Now, some may argue that this activity is just a word exchange; it doesn't really mean anything unless actions are taken. We would definitely agree actions need to be taken. However, it is important that we be able to visualize these statements. The power of visualization of what you want to achieve is crucial to making it happen. Any world-class athlete knows that unless you can visualize it to the very detail, success is unlikely. That is why we must engage in this reframing activity—to clean the air we breathe. We must call out the strength-based descriptors. Then, reflect on why it might be difficult to imagine these descriptors and do the internal antiracist work of coming to terms with why such different actions can be taken.

Taking Action

It is important to tie this reframing work to actions that can change school structures and classroom instruction. Table 5.4 provides a set of strength-based questions that can drive action to support Black boys to thrive—to be Math Strong.

Table 5.4. Strength-based actions

Deficit-based thinking	Strength-based thinking	Action-driven questions
Poor math students	Mathematical thinkers	1. What opportunities have Black boys had to engage with rich, rigorous, and relevant mathematics tasks? 2. How has instruction elicited mathematical thinking from Black boys in the classroom? 3. What kind of meaningful feedback have Black boys received on formative and summative assessments?
Pipeline to prison	College bound	1. How and where are Black boys placed in mathematics classes? 2. Do Black boys have opportunities to participate in college-bound programs (e.g., AVID, Algebra Project) and enrichment STEM-based activities such as robotics and makerspace activities? 3. How do we support the interests and future aspirations of Black boys at our school?
Disruptive/not motivated	Knowledgeable leaders	1. How does our mathematics curriculum connect to the interests, strengths, and lived experiences of our Black boys? 2. What leadership development programs include Black boys? 3. In what ways are mathematical contributions of Black boys promoted in collaborative learning activities?
Families don't care; don't value education	Families care and value education	1. How do we engage Black families as partners for mathematics learning? 2. How do we communicate to Black families about upcoming enrichment learning opportunities/field trips for their child? 3. How do we structure parent-teacher conferences for Black families to share mathematical strengths and areas of growth of their child?

We invite you to engage in this action-oriented activity with your professional learning communities. As part of your school's equity or school improvement plan, such activities can help prioritize positive outcomes for all groups and their mathematics learning experience. The action-driven questions are designed to provoke purposeful action planning that will lead to positive outcomes for students. These questions are tied to school structures and policies, curriculum and instruction, and communication with parents, families, and caregivers. Being accountable for those actions is key because without accountability, the status quo (the polluted air) remains intact.

To begin with, make a commitment to focus on learning and not labeling. Refuse to use deficit-based terms to describe children, their families, and their communities. If deficit-based comments arise in conversation, call it out and use strength-based language to reframe the discussion. This could include using growth mindset language that focuses on mathematical strengths as well as areas of growth (Boaler 2015). But it cannot stop there. It must also include explicit commitments to antiracist equity-focused instruction that includes strong relationships with students and their families. In this book, we offer five equity-based practices that can deepen your mathematics instruction and that affirm students in light of their humanity to be Math Strong.

Conclusion

In this chapter, we have described a concept called Math Strong to help shift educators from deficit-based thinking to strength-based thinking. *This shift will support all children in light of their humanity and ordinary brilliance to be Math Strong.* Mathematical learning needs to be freeing for children to find joy, be curious, and solve complex problems for themselves and their communities. We have shared multiple activities with examples that surface beliefs about working with students. Those beliefs are often steeped in deficit thinking and tied to structural racism and income inequality. Although we are all breathing this polluted air, too many students experience violence in our mathematics classroom through structures such as tracking practices, high-stakes testing regimes, and meaningless curriculum. We can change this reality as teachers, leaders, and researchers. The reframing activities are a start to the dismantling of structures and practices that harm children while learning mathematics. Equity work is hard and transformative work. As educators, we are identity workers and play a crucial role in cultivating and sustaining a positive mathematics identity and the matching learning experiences that come along with it. This includes rethinking our beliefs, structures, and instructional practices that foreground equity and dismantle racism to make mathematics learning more meaningful, humanizing, and just.

DISCUSSION QUESTIONS

1. How do your mathematics curriculum, instruction, and assessment practices cultivate Math Strong students?

2. What are three things you can commit to that will disrupt/interrupt deficit-based beliefs and structures in your school/district setting?

3. What are three indicators of success that make mathematics learning more humanizing and just for the children you teach?

Part 2

Rethinking Equity-Based Practices

Part 2 of this expanded edition highlights five equity-based mathematics teaching practices that strengthen mathematical learning and cultivate positive student mathematical identities:

- *Going deep with mathematics.* Developing deep understanding of mathematics is a major goal of equity-based mathematics teaching practices (Aguirre 2009; Gutstein 2006). Lessons include high cognitive demand tasks that support and strengthen student development of the strands of mathematical proficiency, including conceptual understanding, procedural fluency, and problem solving and reasoning (National Governors Association Center for Best Practices and Council of Chief State School Officers 2010; National Research Council 2001a; Stein et al. 2000).

- *Leveraging multiple mathematical competencies.* Recognizing and positioning students' various mathematical backgrounds and competencies is a key equity-based practice (Featherstone et al. 2011; Horn 2012; Turner et al. 2012). All students have different mathematical strengths that can serve as resources for learning and teaching mathematics.

- *Affirming mathematics learners' identities.* A positive, productive mathematics learner identity contributes to the mathematical learning of a child (Berry 2008; Boaler 2002; Gholsen and Wilkes 2017; Joseph et al. 2019; Langer-Osuna and Esmonde 2017; Martin 2000, 2009a; Stinson 2008). Instruction that values multiple mathematical contributions, provides multiple entry points, and promotes student participation in various ways (teams, groups, and so on) can aid the development of a student's mathematical learning identity.

- *Challenging spaces of marginality.* Traditionally, mathematics learning has been an independent and isolating experience with a focus on lecture and seatwork. Further, students who do not perform well in this traditional classroom setting are often marginalized, ignored, or positioned as "dumb" (Boaler 2002; Jackson 2009). Practices that embrace student competencies, diminish status, and value multiple mathematical contributions are essential to cultivate (Aguirre et al. 2012; Aguirre and Zavala 2013; Featherstone et al. 2011; Goffney and Gutiérrez 2018; Horn 2012; Joseph 2022; Zavala and Aguirre 2023).

- *Drawing on multiple resources of knowledge.* Equity-based teaching depends on the capacity to recognize and intentionally tap students' knowledge and experiences—mathematical, cultural, linguistic, peer, family, and community—as resources for mathematics teaching and learning. Drawing on this knowledge and experience includes helping students bridge everyday experiences to learn mathematics, capitalizing on linguistic resources to support mathematics learning, recognizing family or community mathematical practices to support mathematics learning, and finding ways to help students learn and use mathematics to solve authentic problems that affect their lives (Aguirre 2009; Aguirre et al. 2012; Aguirre and Zavala 2013; Bartell et al. 2019; Brenner and Moschkovich 2002; Civil 2007; Goffney and Gutiérrez 2018; Gutiérrez 2002; Gutstein 2006; Moschkovich 1999; Simic-Mueller, Turner, and Varley 2009; Staats 2009; Turner et al. 2012; Turner and Strawhun 2007; Zavala & Aguirre 2023).

The preceding list of practices is not rank ordered; all five practices are important to use in the classroom, though not always at once or in every situation. The chapters that follow present case vignettes and examples of lessons that span elementary, middle, and high school and serve to bring these equity-based practices to life. Chapter 6 focuses on cultivating mathematical agency by mathematizing a middle school student's claim of racial profiling. Chapter 7 discusses how elementary teachers can build on student strengths, especially for those whose ordinary brilliance is masked or marginalized. Chapter 8 explores varied experiences of Black girls in high school navigating inequities of mathematics education through their own agency as well as family and teacher relationships. Chapter 9 focuses on rethinking assessment through the five equity-based practices including how to provide meaningful feedback and redesigning grading policies and practices to promote learning and positive identity development K–12. Although each of the five equity-based practices is discussed individually, we recognize that in the classroom these practices most often happen simultaneously and in various combinations, and sometimes spontaneously.

We believe that the work of most teachers intrinsically includes elements of each of these five practices. However, tying these practices to mathematical learning and identity may be a new challenge for many. We encourage teachers to take stock of what they offer their students in relation to these practices and to build on those teaching strengths while acknowledging elements that may need additional work and making wise choices about ways to expand their use.

To assist readers with the ideas in part 2, we present a chart that summarizes the five equity-based instructional practices. The chart identifies and describes the characteristics of lessons that represent each practice. It also provides contrasting characteristics of nonrepresentative lessons, along with assessment considerations and companion questions to spark further discussion and self-reflection. In addition, as part of the expanded edition, we provide an instructional analysis tool of the five equity-based practices that uses a 1–5 rubric scale and descriptors to support professional reflection and peer feedback (see appendix).

We believe that these practices are doable. But they are also intentional and complex. Attention to a few of the practices is a good start, and we encourage you to make strategic choices that will lead you to integrate all these practices into your classrooms over time. As in part 1, the names of students, teachers, and schools that appear in the vignettes and examples are pseudonyms unless otherwise noted.

Five equity-based practices in mathematics classrooms

	A representative lesson	A nonrepresentative lesson	Assessment considerations	Questions for reflection
Going deep with mathematics	Supports students in analyzing, comparing, justifying, and proving their solutions. Engages students in frequent debates. Presents tasks that have high cognitive demand and include multiple solution strategies and representations.	Promotes memorization without examination. Encourages students to follow procedures step by step. Presents tasks that have low cognitive demand and emphasize one solution strategy.	A task— • requires demonstration of multiple strategies or representations; • involves analysis and justification. Communication— • offers meaningful feedback that draws students' attention to "making sense" of the mathematics; • focuses on moving students' thinking forward.	How does my lesson promote mathematical analysis? How do I support students in closely examining the math concept?
Leveraging multiple mathematical competencies	Structures student collaboration to use varying math knowledge and skills to solve complex problems. Presents tasks that offer multiple entry points, allowing students with varying skills, knowledge, and levels of confidence to engage with the problem and make valuable contributions.	Promotes individual progress at specific, predetermined levels of ability. Often structures group work by ability. Presents tasks that are rigid and highly sequenced. Requires students to show mastery of skills prior to engaging in more complex problem solving.	Assessing a task— • calls for a diversified rubric and an answer key that includes math practices such as examining patterns, generalizing, abstracting, making comparisons, and specifying conditions; • requires looking for multiple ways that students demonstrate their knowledge, such as through the use of language, gestures, pictures, physical models, and concrete objects.	How do I identify and support mathematical contributions from students with different strengths and levels of confidence?

Five equity-based practices in mathematics classrooms—continued

	A representative lesson	A nonrepresentative lesson	Assessment considerations	Questions for reflection
Affirming mathematics learners' identities	Is structured to promote student persistence and reasoning during problem solving. Encourages students to see themselves as confident problem solvers who can make valuable mathematical contributions. Assumes that mistakes and incorrect answers are sources of learning. Explicitly validates students' knowledge and experiences as math learners. Recognizes mathematical identities as multifaceted, with contributions of various kinds illustrating competence.	Is structured to emphasize speed and competition. Connects mathematical identity solely with correct answers and quickness. Explicitly discourages mistakes and immediately corrects them, often without constructive feedback. Gives ambivalent value to flexibility, reasoning, and persistence.	Communication— • focuses feedback on making sense of the mathematical idea rather than on the ratio of correct answers to the total possible; • focuses on strengths and improvements needed; • points out what is productive or problematic about a student's chosen strategy.	How do I structure my interactions with students to promote persistence with complex math problems? How do I discourage my students from linking speed with math "smartness"?

Five equity-based practices in mathematics classrooms—continued

	A representative lesson	A nonrepresentative lesson	Assessment considerations	Questions for reflection
Challenging spaces of marginality	Centers student authentic experiences and knowledge as legitimate intellectual spaces for investigation of mathematical ideas. Positions students as sources of expertise for solving complex mathematical problems and generating math-based questions to probe a specific issue or situation. Distributes mathematics authority and presents it as interconnected among students, teacher, and text. Encourages student-to-student interaction and broad-based participation.	Disconnects student experiences and knowledge from the mathematics lesson or presupposes that students' knowledge and experiences are inconsequential to learning rigorous mathematics. Ascribes mathematics authority to the teacher or the text. Relegates complex problem solving to the end of lessons or reserves it for "more advanced" students. Segregates specific students (for example, those viewed as "low ability" or labeled as "English language learners") from the main activities. Restricts student "voice" to a few (often privileged) students.	A task— • emphasizes public discussion of mathematical ideas (whole-class, small-group, pair-share); • requires reasoning behind correct and incorrect solutions.	How do I connect my students' knowledge (in school and outside school) with the main math concept of this lesson? How do I structure a task to maximize student-generated math questions? How do I make sure that all students have opportunities to demonstrate their mathematics knowledge during the lesson?

Five equity-based practices in mathematics classrooms—continued

	A representative lesson	A nonrepresentative lesson	Assessment considerations	Questions for reflection
Drawing on multiple resources of knowledge (math, culture, language, family, community)	Makes intentional connections to multiple knowledge resources to support mathematics learning. Uses previous mathematics knowledge as a bridge to promote new mathematics understanding. Taps mathematics knowledge and experiences related to students' culture, community, family, and history as resources. Recognizes and strengthens multiple language forms, including connections between math language and everyday language. Affirms and supports multilingualism.	Treats previous math knowledge as irrelevant or problematic (assuming, for example, "They lack skills," or "They don't know any math"). Builds on negative stereotypes of the culture, community, or family, preventing math lessons that connect with authentic knowledge and experiences of students. (Such negative stereotypes include notions like "Many parents are laborers—they can't help their children with math," "Asian families support mathematics—that's why Asian students are so good and so quiet," and "That is not how we do division in the United States.") Discourages mathematics discourse because it is deemed too difficult for students who have not mastered standard English. Supports English as the only language spoken in the classroom.	A task involves the creation of stories or situations to solve or represent the problem. Communication offers connections to mathematical ideas that students may know but did not use in their solution or explanation.	How do I make connections with students' previous math knowledge? How do I get to know my students' backgrounds and experiences to support math learning in my classroom? How do I affirm some of my students' multilingual abilities to help them learn math? What impact have race and racism had on my mathematics lessons? How can I learn from family and community members to support my students' mathematical confidence and learning? How can I effectively communicate with families the strengths and needs of students to affirm their math identities and promote math learning?

Chapter 6

Cultivating Mathematical Agency: "He Was Suspended for Being Mexican"

An urban middle school mathematics teacher was deeply disturbed by an exchange with one of his seventh-grade students:

> *Mr. C:*   Joaquín, where is Mario today?
>
> *Joaquín:*   He ain't here. He was suspended for being Mexican.
>
> *Mr. C:*   What did you say?
>
> *Joaquín:*   This school is always picking on Mexicans.

Mr. C wondered whether Joaquín's claim could be true. Could the school be engaged in racial profiling of students? What could he do? Through critical reflection, confidence in his instruction, and commitment to his students as strong mathematical learners, this teacher created two standards-based data analysis mathematics lessons that engaged his middle school students in a mathematical investigation to determine whether Joaquín's claim was true.

The equity-based practices in Mr. C's mathematics classroom facilitated deep mathematical analysis (using concepts related to ratio, percent, and proportional reasoning) of an issue that strongly affected his students. Furthermore, the lessons empowered students, giving them a better understanding of how mathematics can be useful beyond the classroom walls and can promote social change. The lessons nurtured positive mathematical identity and collective mathematical agency among Mr. C's students.

The Claim: "He Was Suspended for Being Mexican"

Mr. C, a National Board Certified Teacher, had been teaching in urban elementary and middle schools for more than twelve years. A white male with a strong interest in mathematics, Mr. C had been teaching at this urban middle school for five years. Midway Middle School and its community had experienced a recent demographic shift. The school's longtime African American population had significantly decreased while three new immigrant populations—Cambodian, Mexican, and Ukrainian immigrants—were taking up residence in the nearby public housing development. The school had the highest rates of poverty (90 percent free and reduced-price lunch) and racial diversity (90 percent students of color) in the district. Racial and ethnic tensions were evident among the different populations of the school. Furthermore, because of low test scores, this school was currently in the state's school improvement program and was under a great deal of pressure to improve student achievement, especially in mathematics.

In the face of these extensive challenges and complexities, Mr. C cared deeply about his students and was dedicated to supporting their success. This was an anchoring commitment of his instructional practice. In fact, he routinely posted the following promises in his classroom:

- I will work with you until you understand.

- I will not waste your time—every activity is tied to a learning standard.

- I will ensure that our classroom functions as a positive learning community.

- I am open to suggestions.

- I will learn along with you.

These five promises set the tone for Mr. C's mathematics classroom. Clearly, Mr. C was a teacher who valued learning (his own and his students') and was open to creating a classroom community that was supportive and positive for all.

In this context, Mr. C was seriously troubled by Joaquín's claim about the school "picking on Mexicans." He shared his concerns with one of this book's authors, Julia Aguirre, with whom he had previously collaborated on a mathematics education project. She suggested that he mathematize Joaquín's claim—in other words, use mathematics to determine whether the claim was true or false. Joaquín's statement was a *claim*, she emphasized. What was the evidence? She gave Mr. C a book, *Rethinking Mathematics: Teaching Social Justice by the Numbers*, by Eric Gutstein and Bob Peterson (2005). She called his attention to a short chapter on racial profiling, titled "Driving While Black/ Brown." She suggested that Mr. C read this chapter and see whether he could draw some parallels to help his students analyze this claim about suspensions.

Mathematics Lessons to Evaluate the Claim

The following week, Mr. C contacted Dr. Aguirre to share his excitement about what had transpired in his seventh-grade mathematics class as a result of two lessons that he had designed and implemented. He had read the chapter and created the two lessons to investigate Joaquín's claim (activity sheets for "Midway Suspensions" [lesson 1] and "Two Sides to Every Story" [lesson 2] are available at nctm.org/more4u). Furthermore, he had talked about the situation with his principal, who had agreed to come to the class to hear the students' conclusions. Mr. C reported that the students immediately engaged with this mathematical task, showed sustained persistence in constructing a mathematical argument, and presented their conclusions orally and in writing to the principal. The following discussion outlines the equity-based practices of Mr. C's mathematics lessons.

Going deep with mathematics

Mr. C designed two lessons that required students to determine the validity of Joaquín's statement by organizing and analyzing real data that he obtained on suspensions at Midway Middle School. In the first lesson, "Midway Suspensions," students investigated data on the general population at Midway, sorted by racial and ethnic group, along with data on the numbers of suspensions by students in these groups and additional data on the numbers when multiple suspensions by individuals were taken into account. For the second lesson, "Two Sides to Every Story," students worked with grade-level demographic data (by race/ethnicity and gender identity) on suspensions for particular offenses. Both lessons were tied to specific grade-level state standards on data analysis. The second lesson, for example, supported the process "Solves Problems and Reasons Logically," and more specifically, the student's ability to "draw conclusions and support them using inductive and deductive reasoning."

As students began their work with "Midway Suspensions," the mathematical complexity of the activity became evident in the comparisons of ratios that it required them to make. They increased their computational fluency by representing ratios as fractions, decimals, and percentages. They generated mathematical questions related to other factors that might show the claim to be true or false. This work raised the issue of the "reasons" for suspensions at Midway. Were specific groups of students overrepresented in suspensions for a particular offense?

Mr. C designed the second lesson specifically to enable his students to explore these student-generated mathematical questions. This activity had multiple goals. It facilitated group collaboration and construction of a mathematical argument, and it included a social action component to galvanize change. Figure 6.1 presents this social action component, which proposed a letter-writing campaign using math-based arguments as a plan to make positive change in the school's learning environment.

> If your group had the power to change school policy, given your mathematical analysis, which offenses would you target to change? Why? Write a letter to the principal that describes your plan for positive change in the student learning environment at Midway. (Minimum length is 1 page.)

Fig. 6.1. Mathematizing school policy

These two activities engaged students in complex mathematical problem solving with a specific purpose that sustained interest and sparked additional mathematical questions. The lessons certainly were not the only kind of mathematical activity in this class, but going deep with the mathematics in these lessons resulted in an increased level of mathematical engagement and discourse, with students debating, computing, making mathematical comparisons and justifications, and communicating those ideas orally and in writing to support a position.

Leveraging multiple mathematical competencies

The tasks of Mr. C's lessons had multiple entry points, thus facilitating engagement by students with varying mathematical competencies. The activity sheets structured an explicit process of mathematical investigation. Collaborative teams were a norm in Mr. C's classroom, and the teamwork positioned various students as experts in this process while underscoring the need for engagement and multiple mathematical competencies. For example, in working on the lessons on suspensions, some students with strong computational skills demonstrated to team members how to convert ratios expressed as fractions, obtained through division, into percentages. Other students with a depth of conceptual understanding pointed out which pieces of data needed to be compared and why. Mr. C structured the teams with specific roles and responsibilities that reinforced support, accountability, and progress; the roles are delineated explicitly in the activity sheet for lesson 2, "Two Sides to Every Story." He reported that his students showed a desire to work together and learn from one another. The strategies that he used recognized and leveraged different mathematical competencies to facilitate complex mathematical problem solving.

Affirming mathematics learners' identities

The second lesson, "Two Sides to Every Story," was inspired by the student-generated mathematics questions elicited in the first lesson. Students began to conjecture why members of specific groups of students were being suspended. Was there a relationship between race or ethnicity and specific offenses? Mr. C believed that his students were primed and ready to validate their questions in the second lesson. He observed them poring over the data to answer questions. Mr. C identified this enthusiasm, persistence, and analysis as evidence of his students' expanding positive mathematics learner identities. The students wanted to know more. They wanted to validate their own claims and make their case to the principal.

Although Mr. C's students had no problem sharing their opinions about issues, he believed that this was the first time that they had experienced the need for mathematics to lend support to a position—"to back up their claims." He saw their confidence increase as they prepared their arguments and wrote their letters.

Challenging spaces of marginality

Mr. C guided his students in mathematizing a specific claim that had been made by a student and reflected a larger societal reality experienced by many students of color and their families—racial profiling. These lessons also tapped into an implicit undercurrent of racial tension that had negative effects on student and community relationships. Through the opening written statement of the first lesson (see figure 6.2), Mr. C. made his position about racism clear to his students.

"He was suspended for being Mexican"

Last week, I heard one of my students say this in class when describing why a friend of his was recently suspended. If this is true, I want to address the injustice immediately because I refuse to work in a racist school. Before I complain to our principal, I need to have data ready.

Fig. 6.2. Mr. C's statement on suspensions

In fact, Mr. C set up the lessons in response to racism, modeling for students the need to have "data ready" to "address the injustice" if the allegation proved to be true. This white male teacher openly used the word *racist* to describe the alleged policy, and he explicitly positioned himself against such policies and declared his commitment to changing them. He enlisted his students to help, and they enthusiastically responded. The investigations positioned the students as mathematical experts who could give voice to an authentic concern that they experienced (Turner and Strawhun 2007). The lessons made students problem solvers and advocates for themselves and others, thus *centering*, instead of marginalizing, them as confident mathematical learners with a purpose.

Drawing on multiple resources of knowledge

A strength of Mr. C's suspension lessons was that they drew on the students' resources of knowledge to engage them in complex mathematical problem solving. The context of the lessons was an authentic issue that affected students. Their teacher gave them access to real data that required their mathematical knowledge to organize, analyze, and use to support a position. They drew on their own knowledge to generate additional conjectures about the types of offenses that might be involved. In addition, the students worked in teams, which provided peer and mathematical resources to solve these novel and complex problems. Furthermore, their letters to the principal revealed additional information about peer-to-peer interactions that were fueling negative race relations within the school, and this information prompted surveys and other strategies to promote school-community dialogues. The lessons had a positive impact on mathematical learning and identity, as well as school-community interactions.

Conclusion

Although some of the strategies that Mr. C used were already in his teaching toolbox, these lessons extended mathematical learning and engagement in more substantive ways than Mr. C had succeeded in doing before in his classroom. For example, Mr. C believed that one of his strengths in mathematics lesson design was his ability to connect mathematics to students' lives and interests. He considered this to be a part of the multicultural emphasis in his teaching. He knew that students liked cars, had jobs, or were interested in popular culture and technology, so he often introduced mathematical ideas by using these contexts. Yet, the suspension lessons were different. They were driven by a troubling claim made by a student that related to alleged systemic racism. Although Mr. C was deeply affected by this claim, the thought of using this situation as a mathematical context for a lesson did not occur to him. Dr. Aguirre's suggestion of a related reading helped expand his repertoire of meaningful mathematical contexts:

> I think my practice reflects my tools and understanding I have at the time. When I started teaching, I was content incorporating a pretty surface-level multicultural lens. For example, I had posters up showing math from cultures around the world. Later I started thinking about accessing the diverse knowledge my students bring, but even then, it was mostly about trying to make the content of word problems more meaningful and relevant. But after reading Gutstein's critical pedagogy, I realized how much I needed another tool or model to help me think about making math meaningful and accessible to my students. Reading Gutstein's work helped.

An important dimension of Mr. C's identity as a mathematics teacher was making mathematics meaningful and accessible to his students. As Mr. C suggests, this particular dimension evolved over time from making a more superficial connection (for example, through multicultural posters) to making a more critical connection involving issues of equity and social justice.

Mr. C combined many of his successful instructional strategies—such as tying lessons to learning standards and using teams with accountability systems to monitor progress—with a new mathematical context directly tied to an authentic problem faced by many of his students of color. When asked what made these mathematics lessons different from just "good teaching," Mr. C said that he believed the lessons enabled students to tackle a serious issue that needed to be addressed within the school and community. It gave the students a sense of the importance of using mathematics to support positions instead of just relying on opinion.

Although Mr. C knew that he took a risk in having students analyze real data related to suspensions and race, he had faith that his school's administration had no deliberate intention of engaging in racial profiling and that the principal would be open to listening to students' analyses and ideas. Mr. C's confidence in carrying out the lessons was tied to his skills as a National Board Certified Teacher, his district reputation as a strong mathematics teacher, and a positive relationship with the administration. Further, the lessons embodied Mr. C's commitment to being an "advocate for kids." He was a critical partner in his students' mathematics learning process, and he helped position his students mathematically and socially to investigate a difficult situation. He believed that the lessons had a positive impact on their sense of mathematical identity and agency. The lessons empowered students as mathematical learners, giving them confidence and a sense of purpose as they worked to gain insights and offer solutions to complex problems and make positive changes in their learning environment.

The equity-based practices of Mr. C's mathematics lessons facilitated mathematical learning, positive mathematical identity, and collective mathematical agency for students. Mr. C drew on his existing areas of expertise to create an innovative set of mathematics lessons that validated and strengthened his students' views of themselves as mathematical learners. Valuing his students' learning was already a part of Mr. C's mathematics teacher identity. He was a dedicated advocate for his students. He was committed to high expectations through mathematics standards. And he was devoted to helping students be intellectually and socially responsible for their learning.

Mr. C also faced common political and social challenges of urban schools, including accountability pressures to raise test scores and underlying racial tensions that had an impact on the learning environment of the school. His strong stance on making all lessons standards-based illustrated his commitment to high expectations within a high-stakes accountability system. At Midway, he had curricular flexibility as long as lessons were tied to standards. In addition, with these lessons, he took an explicit, public anti-racist stance. Mr. C made it clear to his students, 90 percent of whom were students of color, that he did not want to work in a "racist school." By acknowledging publicly what the claim alleged—that it charged that the school was racist—he positioned himself as an advocate for social justice in the eyes of his students.

Mr. C encouraged mathematical analysis and agency. His lessons enabled students with various types of mathematical competence to make valuable contributions. The lessons were tied to an authentic situation experienced by many students, whose knowledge and experiences they centered rather than marginalized. Students drew on their own resources—math, peer, community, and so on—to construct solutions and provide teachers and administrators with ideas for change. The lessons tapped into the expertise of students and advanced their mathematics learning. They positioned the students as possessors of mathematical resources and agency for analyzing and solving the problem related to Joaquín's claim.

DISCUSSION QUESTIONS

1. How do your curriculum and instruction cultivate mathematical agency?

2. What similarities and differences do you see between your commitment to student learning of mathematics and Mr. C's commitment? Are the outcomes of student engagement and mathematics learning in your classroom similar to those in Mr. C's classroom?

3. What kinds of authentic problems do your students face? Could you mathematize any of these problems? What are some ways in which you could identify pressing issues that students might want to change? How might you mathematize these problems?

Chapter 7

Building on Students' Strengths: The Case of Curry Green

The teachers at Rosa Parks Elementary School were having their monthly faculty meeting. This week's topic focused on mathematics—specifically, how to increase the mathematics achievement of the "Title I" students at the school—students considered "low performing" by law in schools with relatively high poverty rates, and all students in many of the nation's highest-poverty schools. As Ms. Davis headed back to her fourth-grade classroom, she ran into Mr. Thompson:

Mr. Thompson: I hear you'll be getting Curry Green next week. He's really far behind. He doesn't know any of his multiplication facts. I've given him extra practice sheets for homework. They don't seem to help. He has trouble focusing and distracts the other kids with his drawing. All he wants to do is draw, all the time. It's gotten to be a real problem.

Ms. Davis: I was thinking of starting my stations next week. Maybe I'll include multiplication array flash cards. It just might be what he needs.

Mr. Thompson: Believe me, I know what works. He needs to know his facts. None of these Title I kids know their facts. They'll never pass algebra without understanding fractions, and they can't learn about fractions without knowing their multiplication facts.

Ms. Davis: Knowing their multiplication facts. Is it really that simple?

Mr. Thompson: Yes, it is. There are some skills that are prerequisites for others. Math is a linear path of skills.

The teachers in the preceding dialogue represent two opposite yet common perspectives on the teaching and learning of mathematics. Mr. Thompson believes that learning and teaching mathematics are linear processes and that a teacher's focus should be on helping students achieve mastery of a defined set of mathematical rules and procedures. When students have acquired new knowledge, they proceed to the next set of concepts, with each concept building on the previous one. By contrast, Ms. Davis's comments reflect her belief that a student can learn and understand mathematics in multiple ways if given the opportunity. The Common Core State Standards for Mathematics (National Governors Association Center for Best Practices and Council of Chief State School Officers 2010) emphasize the importance of focusing on both conceptual and procedural knowledge. Mathematical understanding and procedural skills are equally important, and teachers can assess students' progress in developing both by using mathematical tasks of sufficient richness.

The mathematics perspective that teachers embrace has an impact on their view of their role and their effectiveness as educators (teacher identity) and subsequently governs the content that they teach and the instructional practices that they employ. Simply put, what teachers believe is important influences the decisions that they make about what content to teach, how to teach it, and, in many cases, who should receive the content (Stipek et al. 2001). In this way, teaching is no different from many other areas in life. We make decisions on the basis of what we believe.

This chapter explores the impact of the teacher's beliefs on students' learning mathematics and the role that a math teacher has in determining a student's opportunity to learn and succeed. The chapter expands the vignette about the teaching of Curry Green to explore the following teacher beliefs, which in turn shape lesson design:

1. Students should acquire conceptual as well as procedural mathematics knowledge.

2. Students' experiences and prior knowledge can be vehicles for developing conceptual and procedural knowledge.

3. Teachers have a responsibility to design learning experiences that allow each student to feel capable and successful.

Ms. Davis reflected on her mathematics instruction and constructed a lesson that incorporated many of the equity practices described in the introduction to part 2. Keep in mind that these practices often happen simultaneously in a variety of classroom interactions, making it at times challenging to identify and delineate each completely. Here we focus on the following equity-based practices that Ms. Davis used to support Curry:

- Challenging spaces of marginality
- Leveraging multiple mathematical competencies
- Going deep with mathematics
- Affirming mathematics learners' identities

The result was an engaging mathematics environment that supported Curry Green's math identity. Ms. Davis's lesson construction demonstrates how an attribute of a student's identity that has been a problem in one class can become an asset in another.

The vignette illustrates that building on students' strengths can have a positive impact on the development of mathematical identity, learning, and engagement.

Mathematics Learning and Teaching at Rosa Parks Elementary

Curry Green is an African American boy in the fourth grade. He attends Rosa Parks Elementary School. The school is situated in a university town; consequently, it is very diverse ethnically, culturally, and socioeconomically. Some students have parents who are teaching for a year or so at the university, others have parents who work at the neighborhood restaurant, and still others are members of families on public assistance. The school is located in an urban district where more than 65 percent of the students are in the Title I program. The staff has undergone a recent shift, with six teachers transferring to the school from other schools as a result of budget constraints and school closures. Many of these are veteran teachers. This influx has stimulated new conversations among staff members about instructional practice and student achievement.

Although this is Curry's first year at Rosa Parks Elementary School, he already knows the school secretary very well. His teacher, Mr. Thompson, often sends him to the office. Mr. Thompson is white and has been teaching for fifteen years. He taught sixth grade at one of the middle schools in the district but, because of budget cuts, has been reassigned to the elementary level. Mr. Thompson loved mathematics when he was growing up. It was an easy subject for him throughout most of his schooling. He considers himself a good mathematics teacher since most of his sixth-grade students performed well on district benchmark tests. In conversation with Ms. Davis, he expressed frustration that some of the teachers didn't understand what was needed for mathematics success in the upper grades.

Ms. Davis is African American and has been teaching for six years. She regularly participates in some type of mathematics professional development in the summer to strengthen her classroom instruction. A love of mathematics is not what drew Ms. Davis to teaching. She enjoyed her school years, but she has many friends who don't feel that they did. Ms. Davis became a teacher because she wanted to be able to help other children love school and learning. It bothers her that so many of the African American and Latinx students at her school struggle with math. She hasn't figured out why this is so, but she is determined to keep searching for ways to help her students achieve and be successful. In a conversation with a colleague during a professional development session, she reflected on the origin of her beliefs about mathematics:

I took just enough math to get through college and my credential program. I did well enough but by no means considered myself a math person. It wasn't until we did the square numbers activity that I realized mathematics could be understood. Who knew a square number looked like a square when drawn? Mr. Thompson and I have discussed the best way for students to learn at many faculty meetings. Yes, they do need to know their facts but timed tests and forty-problem worksheets don't work for all kids. I used to assign worksheets with problems for homework or extra practice. I got seduced into thinking since many of my children could compute, they understood multiplication. Some of my students would enter the classroom knowing their facts but didn't know how to begin to solve a multiplication application problem.

Like most teachers, Ms. Davis has seen her understanding of how to teach mathematics evolve over time. Her initial perception of what students need for success and what requirements this need imposes on instruction has also changed. She believes that all students can be successful in mathematics and that it is a teacher's responsibility to provide avenues to engage their thinking. These beliefs had a direct impact on how Ms. Davis welcomed Curry Green into her classroom, built on his strengths, and engaged him in mathematical learning.

Embracing Curry Green

Ms. Davis thought carefully about how to engage Curry Green and instill confidence in him about his ability to think and reason mathematically. Further, she designed her lessons to provide opportunities for others in the class to see Curry as a mathematically proficient student. To achieve her goals, Ms. Davis interwove and blended a number of equity-based practices.

Challenging spaces of marginality

To address Curry's expected anxiety about entering a new class belatedly in the school year, Ms. Davis provided him with a "buddy"—a partner to help him with the class routine and ease his transition to her class. She also arranged for him to be interviewed by a student and his answers to be posted on the class bulletin board. Ms. Davis has designed several welcome activities for new students. She wants each child to feel that he or she is a valued member of the classroom, so she provides opportunities for students to share their uniqueness with their peers. She believes that addressing the affective domain of students' learning is critical to their ability to gain access to higher levels of cognitive knowledge, particularly in the case of students who have not been successful in school. Addressing a student's feelings and attitude about learning is paramount in gaining access to cognitive domain knowledge such as mathematics conceptual understanding (Blum-Anderson 1992).

After recess, Ms. Davis began her math lesson with a problem. She placed a coordinate graph activity (see figure 7.1) on the document reader in her classroom and directed her students, "Take out your math journals."

"What's a math journal?" Curry asked. His buddy explained, "It's like our own workbook without the printed problems." Ms. Davis set the students to work on the task: "OK. Class, you have about a minute to think about the problem by yourself. When the chime rings, you can begin to talk about the problem with your partner." Ms. Davis set her phone alarm for one minute. The children copied the graph into their journals and began to write notes on the pages.

Curry didn't understand. "There aren't any numbers on this graph," he whispered to his buddy. His buddy responded, "We had graphs like this last week. The lines don't have to go by ones. You can skip numbers. They just have to keep the same pattern all the way through the graph."

The alarm chimed, and Ms. Davis asked the class, "What is the problem asking you to do?" She then asked, "How do we figure out what point A represents? Talk with your partner and share your thoughts." The students began to talk to one another. Curry was amazed that they were allowed to talk and work with other classmates. In his other classes, he often got in trouble for talking.

DOT AND RECTANGLE

The dot (Point A) on this graph represents a rectangle whose area is 24 square inches.

Mark two other points on the graph that represent other rectangles with area of 24 square inches.

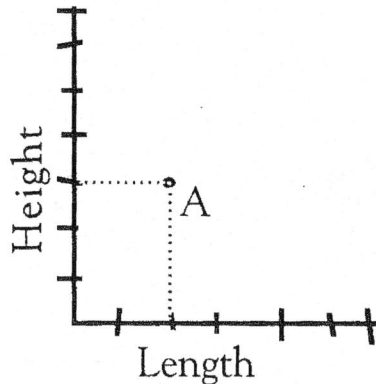

Explain why you put each dot at a particular point.
(Hint: The marks on the lines are not units of 1.)

Fig. 7.1. A coordinate graph activity presented by Ms. Davis

After his buddy explained that each segment of the graph is a unit of 2, Curry jumped in and said, "I see. It's a rectangle, 6 by 4." Curry and his buddy located three more rectangles in the graph.

When Ms. Davis asked for volunteers to show their solutions, Curry was nervous about going up to the front of the room, but he realized that students didn't have to go forward alone. They presented in pairs, and Ms. Davis let each team decide who would speak. No one was put on the spot to explain their work. His anxiety level immediately went down, and he was able to listen to the team present their solutions.

Ms. Davis thanked the presenting teams and posted their solutions on the whiteboard. She said, "We are going on to another activity now. I want everyone to think about whether there's another solution we missed. We'll talk more tomorrow."

Ms. Davis used a common lesson structure, beginning with a math warm-up activity. Many teachers use this technique as a management tool, occupying students with an activity or problem while they take care of routine tasks like homework or attendance. However, a warm-up can also provide students with an opportunity to tap into prior knowledge or experiences before a new concept is introduced. A topic that students are familiar with, like coordinate graphing, provides an avenue to a new application, such as the use of an area model for multiplication. Moreover, Ms. Davis engineered opportunities for Curry to talk with a partner to verify his understanding of the mathematics in the task, a critical element in his engagement in the lesson as a new student in the class. Ms. Davis's beliefs about the benefits of student discourse contrast with the views of Mr. Thompson, in whose class Curry was made to feel marginalized because of his need to talk about mathematics.

Leveraging multiple mathematical competencies

After wrapping up the discussion of the initial problem, Ms. Davis explained the day's task: "Each of the stations today will focus on a key element for understanding and applying what we know about multiplication." Figure 7.2 shows the station activities that Ms. Davis designed to enable her students to see multiplication in different contexts (the lesson is also available at nctm.org/more4u).

Multiplication Stations		
Station 1	**Station 2**	**Station 3**
Multiplication Arrays	Why does it look like a square?	Division: A rectangle, then a little more…
[*Students create a set of array cards.*]	[*Students build, record, and cut out models of the following facts on colored paper: 2 × 2, 3 × 3, 4 × 4, 4 × 5, 5 × 5, … 12 × 12. Then they arrange them in ways that show an increase or decrease in area.*]	[*Students create a rectangular array for each of the following division problems: 15/4, 64/3, 84/5; 65/7; 44/9. Then they represent each one on graph paper.*]

Fig. 7.2. Ms. Davis's multiplication stations

Ms. Davis used multiple representations to assist students in developing a conceptual understanding of multiplication. The sequence of tasks connected concrete, pictorial, and abstract representations of multiplication facts, allowing students to compare rectangular and square arrays and see how division with remainders compares with multiplication arrays. According to the National Research Council (2001b), "Knowledge that is taught in a variety of contexts is more likely to support flexible transfer than knowledge that is taught in a single context" (p. 17). Each station highlighted pictorial representations of multiplication (arrays) since drawing was one of Curry's strengths that Ms. Davis wanted to leverage. This design also provided the opportunity for students who had developed fluency with their facts to go deeper by exploring pictorially the inverse relationship between multiplication and division.

Going deep with mathematics

Ms. Davis called Curry's table to station 1. She explained the task: "You are going to create a set of array cards. They will help you to visualize what your multiplication facts look like and what the fact means." She asked the group, "What does 3 times 4 mean?" Carlos, Curry's buddy, answered, "Three groups of 4." Madison, a girl in the group, said, "No it doesn't. It means 12." Ms. Davis replied, "The total number of items is 12. But let's look at what it looks like."

Ms. Davis took unit cubes and built an array that had a width of 3 and a height of 4. Janelle responded, "But that also looks like 4 groups of 3 if you look at it this way." Ms. Davis replied, "You are correct, Janelle. And Carlos, you were correct, too. This multiplication array, or rectangle, can be represented in two ways. We refer to that fact as the *commutative property of multiplication*. The order of the two numbers doesn't affect their product or answer. Let's draw them on our index cards."

Curry sketched his arrays with a perspective view of the cubes. Ms. Davis looked over in amazement at Curry's drawing. He had drawn a three-dimensional representation of the array. When he saw her facial expression, he quickly erased it and said, "I'm sorry." Ms. Davis replied, "Curry. No. Don't erase it. That's exactly how it looks from an overhead perspective. I am going to draw a two-dimensional representation because it's easier for me to draw."

Ms. Davis stated, "We will also write the related division equation. Who knows what that would be?" Curry looked at the array for a minute and replied quietly, "Twelve divided by 4 equals 3?" Ms. Davis affirmed his answer, "Yes, that is correct." Curry smiled. Ms. Davis followed up with another question, "What's another division equation?" Curry responded tentatively, "Twelve divided by 3 equals 4?" Ms. Davis affirmed his response again: "Excellent! So our array gives us information about both the multiplication and division fact. Let's try a few more." Figure 7.3 illustrates the students' work with Ms. Davis.

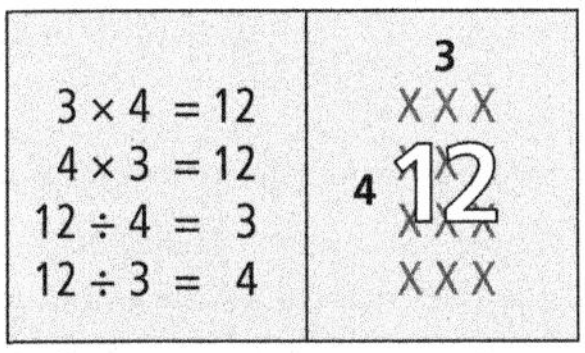

Fig. 7.3. Multiplication and division activity in Ms. Davis's class

Station 1 allowed multiple mathematical competencies to be showcased. The students offered two responses to Ms. Davis's initial question, "What does 3 times 4 mean?" Ms. Davis validated both responses and capitalized on the opportunity to show how they were related through multiple representations, the commutative property of multiplication, and the inverse relationship between multiplication and division in fact families. Curry's participation demonstrated that he was making important learning connections. He was also able to recognize and recall multiplication and division facts while working on this task—a competency that he had been unable to demonstrate in his previous class.

Affirming mathematics learners' identities

Ms. Davis's students continued to work with their partners to create sets of array cards for the 8 and 9 times tables. Curry was surprised when the bell rang for lunch. On the way out, Ms. Davis asked Curry how he had liked working in stations that morning. She acknowledged how challenging making the transition to a new classroom can be and told him he had worked well with his partner in completing the cards. She said, "I know it can be tough walking into the middle of a math lesson in a new class."

Curry responded, "It was OK. It didn't seem like math class to me. It's really different than in Mr. Thompson's room. We get to talk *and* draw." Ms. Davis explained that learning to visualize problems was a mathematics problem-solving skill that everyone needed to acquire. "Some students naturally process information that way, but for others, it is helpful to talk through their math problems to understand what's needed for a solution. There isn't a right or wrong way to think, just different ways." Curry thought for a minute and then said, "I always thought I was the dumb one in class. I just didn't get it sometimes, especially in math. It's so hard to remember all those numbers and rules." Ms. Davis agreed. "It's easier to memorize when you understand what you're memorizing. That's why we always strive for understanding in our class, Curry. It makes us much better mathematicians." Curry asked, "Does that mean no timed tests?" She responded, "No. We all learn at different paces. Speed will come in time. Don't worry." Curry smiled and said, "Thanks, Ms. Davis."

This poignant exchange between Curry and Ms. Davis emphasizes how the mathematics identities of students can be shaped and reshaped through positive mathematics learning experiences. Ms. Davis's responses to Curry affirmed his developing positive mathematical identity. The view that she provided of what it means to learn and be successful in mathematics was broader than the perspectives that Curry had previously encountered. Ms. Davis helped Curry readjust his perception of what it means to be "good at math" and gave him a new vision that included a path potentially leading to both conceptual understanding and procedural skill.

Conclusion

Many third- and fourth-grade teachers regard mastery of multiplication facts as a curricular cornerstone for their students. However, for students, mastering the facts and conceptually understanding what multiplication means are two different things. It is critical to develop proficiency in both. This can be particularly true for students of color, who historically have had limited access to cognitively demanding tasks and courses. Focusing on timed mastery can lead to negative dispositions toward mathematics (National Research Council 2001b).

Mr. Thompson and Ms. Davis, the two teachers portrayed in this chapter, display very different perspectives on mathematics proficiency and the knowledge and skills necessary to achieve mathematics success. Their differences had a significant impact on Curry Green's mathematical experience and his developing mathematical identity. Mr. Thompson's personal experiences as a mathematics learner and educator shaped his opinion of the criteria for math success and his views regarding teaching. When assigned to a new school and grade level, he designed his instruction on the basis of these beliefs and became frustrated because he was not able to achieve the same level of success in his new surroundings that he had attained in his old ones. When teachers inherit a class of students who struggle with mathematics or have been identified as low achievers, too often, they assume that procedural, low-level remediation is most appropriate. Mr. Thompson quickly began to look at what his students *lacked* and what changes they needed to make rather than reflect on his *own* beliefs about mathematics learning and teaching.

Ms. Davis, on the other hand, was guided in her teaching by a belief that learning mathematics is a sense-making process and that the teacher must create a classroom environment where children feel included, and their uniqueness is honored. Curry Green was a student not unlike many students in our schools today. He was a bright African American boy who had not experienced a great deal of success in mathematics. Ms. Davis embraced the multiple identities that he brought to her classroom. She embedded equitable teaching practices in the lessons that she designed to ensure that those lessons would tap into her students' mathematical strengths and enable them to see themselves as successful mathematics learners.

DISCUSSION QUESTIONS

1. Ms. Davis's and Mr. Thompson's professional and personal experiences shaped their conflicting perspectives and beliefs about the necessary elements for success in mathematics. They also had different views of Curry Green as a mathematics learner. Reflecting on your own experiences, what do you believe are the critical components for mathematics success? How are these reflected in your instructional practice? Do Ms. Davis's views or Mr. Thompson's views resonate with your own beliefs and experiences?

2. How might you support students like Curry, who might have a negative mathematical identity but nevertheless have positive identities in other areas of their life?

3. Ms. Davis leveraged Curry's strengths as an artist to deepen his understanding of multiplication and fact fluency. In what ways have your math lessons tapped into a student's strengths?

4. Statistics show that students of color, multilingual learners, and learners from families affected by poverty often receive mathematics instruction that is predominantly procedural and skill-based. What conversations have you had with colleagues at your school site on differentiated instruction for specific groups of students?

Chapter 8

Nothing to Prove: Seeing Brilliance and Creating Access for Black Girls in High School Mathematics

Black girls are especially vulnerable to inequities that exist within mathematics education (Gholson 2016; Gholson and Martin 2014, 2019; Jones 2003).

> As a discipline, mathematics is particularly susceptible to inequities in the classroom because it is shrouded in a myth of objectivity. Indeed, the mathematics classroom is one of many schooling spaces where Black girls suffer, are taken for granted, and dehumanized. (Joseph, Hailu, and Matthews 2019)

The intersectional identities of Black girls, set within the historical foundation of slavery and patriarchy, create fertile grounds for the biases and structural barriers inherent in mathematics education, like tracking to manifest themselves. Mathematics has often been viewed as an objective science, from the focus on calculations to the precise grade books created and maintained by math teachers at the elementary and secondary levels. What we have learned, however, is that behind the facade of objectivity, there is a system and set of ideologies that favors white males not only in the ways of interacting, thinking, and learning but also in the way in which mathematics itself is perceived (Campbell 2012; Gholsen and Wilkes 2017; Joseph, Hailu, and Matthews 2019), hence the questions presented in Chapter 1: What mathematics? For whom? For what purposes? The question as to whether mathematics is necessary for all students, particularly advanced courses, is still debated by many in our education system. The outgrowths of this mindset can be seen throughout math policies, structures, and environments that impede many students of color, especially many Black girls (Joseph 2022).

Despite the incremental outcomes of standards-based reform efforts, the system of mathematics education ultimately self-corrects to recreate hierarchical structures that maintain disparate experiences for the same groups of students. This self-correcting tendency is consequential for Black girls. Although efforts have been made to change instruction, curriculum, and assessments, Black girls' brilliance, logic, intuitiveness, and problem-solving attributes often are overshadowed by descriptors that are labeled as negative behaviors. These behaviors are seen as contradictory to those of the stereotypical "good" mathematics students (i.e., focused, serious, efficient), instead of being viewed as different and complementary expressions of strong mathematical thinkers (Joseph 2022).

In a recent professional development workshop conducted by Karen Mayfield-Ingram, participants engaged in the "What Have You Heard?" activity described in Chapter 5. Mathematics teacher leaders and educators were asked to write on sticky notes what they had heard about various demographic groups of students. When responding to "What have you heard about Black girls?," 20 percent of the descriptors were categorized as positive, 75 percent as negative, and 5 percent as both. Table 8.1 reflects some of those comments.

Table 8.1. What have you heard about Black girls?

Positive	Negative	Both
• Beautiful • Resilient! • Determined	• Disrespectful • Have attitude • Loud • Argumentative	• Strong-willed • Tough

These stereotypical descriptors circulate throughout many of the educational environments where Black girls learn, socialize, and are assessed. They influence how adults perceive them, how other students interact with them, what courses they are allowed to take, the pedagogical tools teachers employ, and the perceptions Black girls hold about their own identities and attributes (Annamma et al. 2016; Epstein, Blake, and González 2017; Joseph 2022; Morris 2016).

The goals of this chapter are to understand how these descriptors and stereotypes are used to position Black girls in their mathematics classrooms. We offer two vignettes from high school contexts and examine equity-based strategies that can counter these descriptors and stereotypes and that reposition Black girls for success. The focus on high school allows insights into the interplay of school policies, infrastructures, and teacher mindsets that influence Black girls' mathematics learning. As you study the vignettes, it is important to note that Black girls are not a monolith. These two examples cannot capture the experiences of every Black girl. However, the examples do highlight what can happen to Black girls. We encourage you to think, beyond stereotypes, about the different ways that Black girls can and do embody and actualize their Black girl identities and how you might respond to their individual and collective needs.

Sydney's Story

Sydney Hamilton is a Black girl in the ninth grade who attends a large public high school in an urban school district. Sydney is an active teenager who participates in martial arts and likes listening to music and being with her family. At school, she is a quiet student who does well in most of her classes except for math. She is very observant and direct in her questions and responses to others. She has been labeled as rude or having "an attitude" by some of her teachers because she generally does not see the need to expand or make things more complex than necessary and tends to question the reason for requirements or directions given.

Math class

In her algebra class, Ms. Linton uses participation sticks to call on students. She called Sydney to come up and share her beginning work on one of the homework problems. Ms. Linton saw Sydney start to work, but she quickly covered her paper.

> *Janelle had a bag of marbles. She gave one-third of them to Rebecca, and then one-fourth of the remaining beads to her brother John. Janelle then had 24 marbles left in the bag. How many marbles did Janelle have in her bag in the beginning?*

Ms. Linton: "Sydney, would you like to come up and show us your work?" Sydney replied, "I'll pass," and turned her head in the opposite direction. Ms. Linton called on another student, who shared an algebraic method for solving the problem. The student made a minor calculation error. Ms. Linton corrected his error without making too much of it and began to set up the work for the day. Sydney mumbled, not so quietly. "That's a lot of extra work." The class laughed. Ms. Linton looked sternly at Sydney and reprimanded her for the remark. Ms. Linton replied, "It's not extra work; it is *the* work. That's what we are here to do, to learn, Sydney. Not to criticize people."

Sydney balled up her paper, tossed it toward the trash, and put her head down on the desk. Ms. Linton sighed, and she, as well as Sydney's table group members, ignored Sydney for the remaining class period. The bell rang. Sydney packed up and walked out. Ms. Linton was very frustrated with Sydney. She noted that Sydney seldom participated and was often rude to her classmates. As she walked to the door, she noticed the paper and was curious about what Sydney wrote, so she opened it. Sydney created a graphic representation of the various steps in the problem (figure 8.1). She divided a rectangle into thirds, crossed out a third, then divided it into fourths, and crossed off one section. She then took the remaining total amount and figured out what each piece was worth to calculate how many marbles Janelle started with. Ms. Linton said to herself, "Hmm. Am I thinking about this all wrong?"

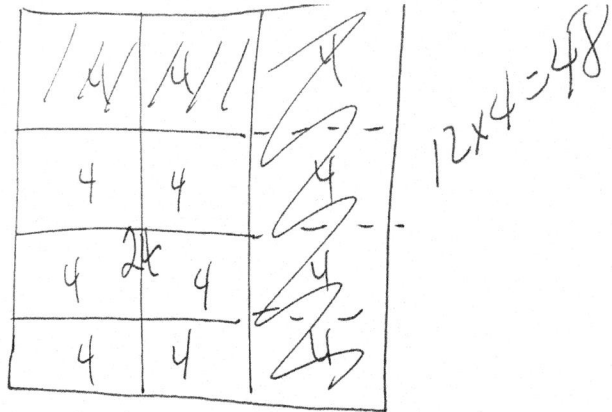

Fig. 8.1. Sydney's math work

Faculty room

Mr. Alston, the history teacher, and Ms. Linton were sitting at a table eating lunch. Mr. Alston noticed Ms. Linton looking at Sydney's wrinkled diagram. He asked, "What's that?" She responded, "A student threw her work in the trash on her way out." "Was it that bad?" he asked. Ms. Linton replied, "No. On the contrary, it's kind of an elegant way of looking at the problem." She relayed the day's interaction with Sydney to Mr. Alston. She said, "I just don't know how to connect with her. How do we adjust midstream?" Mr. Alston responded, "Have you talked with her? Maybe get to know more about her? What class is like from her perspective? It may give you some insight on how to adjust things on your end." "You are right. I'll try to connect with her today after school. Thanks for your insights," replied Ms. Linton.

After school

Sydney walked into algebra class to see Ms. Linton. She was direct. "I need to pick my brother up from soccer practice." Ms. Linton replied, "This won't take too long. Please sit down. I wanted to talk with you about your work." Ms. Linton showed the paper to Sydney. She shrugged. Ms. Linton said, "I think we got off to the wrong start. Obviously, you are a very smart and insightful young woman. But I have a feeling that you are not happy being in class. Do you like math, Sydney?" "It's OK," Sydney said quietly. Ms. Linton asked, "What's the best part?" Sydney replied, "I don't know." Trying again, Ms. Linton asked, "OK. What about what you don't like?" "Homework." They both laughed. Ms. Linton continued, "Is homework difficult for you?" Sydney answered, "No, it's just another thing to do when I get home." Ms. Linton responded, "I get that. I'm curious, what other things do you do at home?" They talked a bit more about Sydney's responsibilities at home. Ms. Linton shared that she also took care of her siblings after school when she was Sydney's age.

Ms. Linton brought out Sydney's paper and asked, "Can you talk to me about your drawing? Maybe re-create it for me?" Sydney pointed to her drawing. "Well, I figured this was Janelle's bag of marbles. If I take away $1/3$, $2/3$ is left. Then if I divide the rest into fourths, then take away $1/4$ for John, the rest is 24. There are 6 sections, so there must be 4 marbles in each. If you divide the whole piece into fourths, you have $3 \times 4 = 12$. And $12 \times 4 = 48$." Ms. Linton stated, "That is a very elegant and efficient solution. Would it be OK to share this with the class tomorrow? I think it would help other students think differently." "Yeah, OK. I gotta go, Ms. Linton," replied Sydney. She gathered her things and left.

Next day

After the warm-up, Ms. Linton began the lesson by talking about the importance of knowing more than one way to solve a problem. She noted, "It will help you develop into stronger math students. Let's take another look at the marbles problem. We had several elegant solutions to the problem." Ms. Linton wrote on the board the algebraic strategy they discussed in class the previous day. Then she asked Sydney if she would like to share the way she solved the problem. Sydney responded, "No, I'm good." Ms. Linton replied, "How about I draw it, and you talk me through it?" Sydney agreed. After talking through Sydney's solution, several students commented, "That's cool. I never thought about it that way." Ms. Linton clarified, "The real power comes from understanding the problem deep enough to decide which method in your strategy toolkit to use. Understanding multiple ways is where the real strength comes from and what's important to solve real problems outside our classroom textbook."

Seeing Brilliance with Equity-Based Practices

Going deep with the mathematics

Ms. Linton's decision to ask her students to show their work in multiple ways is an example of the equity-based strategy *going deep with the mathematics*. Tasks with high cognitive demand allow for different strategies and different understandings and knowledge to emerge. This is particularly important for Black girls who often do not have the opportunity to learn deep, rigorous mathematics or demonstrate their thinking and competencies (Morton and Smith-Mutegi 2018).

Because Sydney was perceived as unengaged and her comments interpreted as unproductive, she initially received a reprimand from Ms. Linton. Sydney's decision to exercise her agency by questioning the efficiency of the algebraic method of solving the problem demonstrates her conceptual understanding of the problem and her willingness to engage in mathematical discourse. Research demonstrates that when Black girls are candid in their responses, they are seen as disruptors instead of displaying Standards of Mathematical Practice such as constructing viable arguments and critiquing the reasoning of others (Archer-Banks and Behar-Horenstein 2012). By explaining to the class that representing solutions to problems in multiple ways would strengthen students' understanding and provide more flexibility for solving problems in general, Ms. Linton not only helped them develop more tools to utilize in subsequent math problems but also expanded their perceptions of what it means to do mathematics and to be a mathematical thinker.

Affirming mathematics learners' identities

When Ms. Linton reprimanded Sydney in front of the class for her comment about the algebraic solution being a lot of extra work, she not only may have affected Sydney's perception of herself as a mathematical thinker, but also may have influenced her classmates' perceptions of Sydney's math abilities because her group ignored her for the rest of the class period. Sydney correctly surmised that her way of thinking about the problem was not valued even though she had a reasonable response to the situation. She believed the class and her teacher discarded her work. Ms. Linton's remark also reinforced the belief that Black girls are "loud," "rude," and "inappropriate" (Morris 2007). Because their responses are frequently characterized in this manner, it can be challenging for many Black girls to develop strong, positive math identities. As teachers, we may not grasp how mathematical discourse changes when Black girls are in the classroom (Joseph, Hailu, and Matthews 2019; Gholson and Wilkes 2017; Gholsen and Martin 2019; Nasir 2011). This is not to say that there is only one way that Black girls should engage in mathematical discourse, but it is important to understand how dominant or preferred modes of discourse can exclude Black girls or make it more difficult for them to engage.

When Ms. Linton followed her curiosity about Sydney's engagement and looked at her work, it caused her to question her mindset about Sydney's math engagement and abilities. Following Mr. Alston's advice and taking the time to talk with her, to get to know a bit more about Sydney and her experience in class, was a first step toward rehumanizing the interaction with Sydney and positively supporting her math identity (Goffney and Gutiérrez 2018). This is what Natalie King (2022) refers to as *spirit healing* for Black girls, attending to their mental well-being and acknowledging the trauma, while unintentional, done to their spirit.

Challenging the spaces of marginality

After getting to know Sydney and listening to Sydney's explanation, Ms. Linton realized there was a missed opportunity to position Sydney as a resource for the class. To rectify this situation, Ms. Linton positioned Sydney's work as an example of another "elegant" representation of the solution. By highlighting to the class its value in the development of a strong mathematics foundation, Ms. Linton validated Sydney's knowledge and strengthened her mathematics identity. In addition, Ms. Linton did not give up on Sydney when she stepped away from the initial invitation to present her work to the class. By offering to reconstruct Sydney's work, Ms. Linton met her halfway, elevating Sydney's voice as a mathematical authority who the class could learn from. It brought Sydney from the margins to the center of the classroom's intellectual space—influencing her willingness to exercise her individual math agency and helping to build the collective agency of the classroom.

Because Black girls face bias within societal, educational, and mathematics contexts, it is imperative to combat the harmful stereotypes that characterize their interactions. It is important to assume their brilliance and intuitiveness and reach out to get to know each girl. That outreach greatly contributes to Black girls' math identity development and creates conditions by which they feel they can bring their multiple identities into the learning environment (Gholsen and Martin 2019).

Elizabeth's Story

Math class

Elizabeth Mack is a Black girl who attends a small high school in a Midwestern suburban city. Elizabeth has always done well in math. She did well in algebra as a freshman and this year was placed in the honors geometry class as a sophomore. On the first day of class Mr. Royal drew points on the whiteboard, connected them, and said, "This is a plane." Elizabeth was confused. It was a picture of a rectangle (not an airplane), but she started taking notes and hoped the class would end soon. She glanced around toward her girlfriends in class. Elizabeth and her three friends were the only girls of color in a class of twenty-five students. The girls made eye contact and shrugged their shoulders, expressing their confusion. The rest of the class seemed to be following along easily. Toward the end of the class, Mr. Royal discussed the course syllabus. "Because this is an honors course, the pace will be much faster than the other courses you probably have experienced. You will be expected to study on your own quite a bit. This course and the paths they lead to aren't suited for everyone, so consider whether you are up for the challenge."

Conversation with parent

A couple weeks into the class, Elizabeth talked with her mom about dropping the math class. Elizabeth said, "I don't think I belong there. I just need more time to process what he's saying. I take good notes and I can memorize stuff, but they are just words." Her mom asked, "Do you ask questions in class?" Elizabeth replied, "Sometimes, but I feel like I'm slowing everyone down. Mr. Royal repeats what he says but it doesn't really help." Mom asked, "Have you talked with your teacher?" Elizabeth responded, "Yeah. We went over my grades on the last two quizzes, $36/48$ and $43/60$. He asked how I studied. I showed him my notebook. He was impressed with my note-taking skills." Her mom interjected, "You do take very organized notes." Elizabeth replied, "He also said maybe I should think about dropping the class if it's too challenging and I'd probably get better grades in the regular geometry class. The pace of the next set of courses will not be any slower. It will be even harder for me, and they are probably not going to be necessary to get into the colleges I would attend."

Her mom replied, "First of all, you don't know which colleges you will attend. Your counselor said there were many scholarships available for first-generation college students. But we'll start talking about that more next year. You are a smart girl, Elizabeth. You wouldn't have been recommended for the course if you weren't capable." Then her mom asked, "How are your friends doing in the class?" Elizabeth responded, "There are only four of us in second period. Monica doesn't understand most of the time either, but her older brother took the class and helps her at home. Stacey and Adriane are thinking about dropping the course too." Her mom said, "OK. Let's wait before we make any changes. Remember what we learned at that Family Math night in middle school about the number of math courses and options?" Elizabeth replied, "Yeah, but I don't want to be an engineer or doctor." Her mom said, "That may be, but you want to have options and not be limited because you don't have enough coursework."

Meeting with the counselor

The next day Mrs. Mack, Elizabeth's mom, made an appointment to talk with the academic counselor. She started the conversation, "I was talking with Elizabeth, and she is having difficulty in her honors geometry class. She wants to drop the class. Apparently, there are three girls who also are planning on transferring out." The counselor was surprised. She asked, "What is their concern?" Mrs. Mack relayed what Elizabeth told her about her experience. After listening, the counselor suggested they wait to make any class changes. She told Mrs. Mack that she would get back to her about a recommendation after she talks with Elizabeth and the math teacher.

Meeting with the teacher

After talking with Elizabeth, Adriane, and Stacey, and reviewing the girls' previous grades, the counselor met with Mr. Royal. "Elizabeth's mom came to see me yesterday. She mentioned that Elizabeth was thinking of dropping the class and so were three other girls." Mr. Royal replied, "Really, I've never met Elizabeth's mother. If she had come to back-to-school night, we could have talked then." The counselor adds, "Elizabeth's mom works nights." The counselor told Mr. Royal that she was concerned that four Black girls, all of whom easily met the prerequisites for the course, are thinking of dropping out of the honors track.

Mr. Royal replied, "Oh. I am surprised to hear that all of them plan to drop the class." Mr. Royal also explained that he felt he was in a tough spot because he needed to cover the department's prescribed syllabus. He felt he had no control over the pacing. He continued by emphasizing that perhaps the students weren't prepared for advanced coursework. The counselor replied, "As I mentioned, I reviewed their grades and teacher recommendations. They met the standard. So how can we support their learning? There seems to be some disconnect between the content and the students. These girls are hardworking, bright, and capable. And I will be honest with you, I am concerned that our advanced courses aren't as diverse as our school population. I was hoping that your math department could help us think through this issue because it's also occurring in other STEM areas."

Addressing the concerns

Mr. Royal brought the issue up at the math department meeting during their on-site professional development day the following week. The teachers discussed the issue of pacing and instructional pedagogy. A few of the teachers also expressed dissatisfaction with the syllabus pacing. One of the math teachers commented, "I don't feel confident that we are building a solid conceptual foundation for some of the topics." They brainstormed strategies that might address the concern and agreed to try a few strategies in their courses.

Mr. Royal decided to implement two strategies. The first was to take class time to incorporate some of the construction and modeling examples in the textbook. He had planned to use them as extra credit, but decided to incorporate them into the beginning of units so everyone would have the experience. He also created interactive math journals where students could reflect on their understanding or questions with him at the end of each lesson.

Mr. Royal also talked with Elizabeth and the other girls. He asked them to hold off switching classes. He said, "I talked with your counselor and we both are in agreement that you belong in this class. As teachers, we are also concerned with the pacing of advanced math classes. We think slowing down and going deeper will benefit everyone. I also want to say thank you for sharing your perspective with your counselor. Your leadership actions will help us be better. Since your mom works nights, Elizabeth, I'm going to give her a call during my prep period to discuss the changes."

Creating Access with Equity-Based Practices

Affirming mathematics learners' identities

In addition to teachers, parents and caregivers are instrumental in affirming math learner identities. Mrs. Mack counters myths about Black parents' involvement or lack of interest in their children's education. Although she doesn't have the experience of attending college herself, she is quite clear and adamant about Elizabeth's future as a college graduate. Black parental involvement has been shown to positively influence Black girls' participation and persistence in mathematics (Howard and Joseph 2022; Howard 2020; Howard et al. 2023). Mrs. Mack seeks out information about math options and directly counters her daughter's self-doubt. She promotes Elizabeth's positive math identity by reminding her of her many attributes that would enable her success. She also asserts her agency as a parent by contacting the counselor and relaying not only her daughter's experience but the developing trend with other Black girls in the class.

Black parents have too often been characterized as noninterested and difficult to connect with regarding their children's education. These characterizations stem from comparisons to the parental practices of many white, middle-class parents (McGee and Spencer 2015; Jackson and Remillard 2005). These practices are normalized by school officials. However, Black parents may not show up in the ways prescribed or expected by school officials—namely attending PTA meetings, school socials, or back-to-school nights—but that does not mean they are not supportive of their children's education. Black parents often recognize the societal and educational threats to their children's self-esteem and general welfare and are even more invested in their children's success (McGee and Spencer 2015; McGee and Pearman 2015; Martin 2006; Washington 2019).

Challenging spaces of marginality

Counselors also play a pivotal role in the placement and retention of students in advanced math courses. Black girls often choose to leave advanced math courses due to feelings of exclusion and deficit views of their intelligence (Morton, McMillan, and Harrison-Jones 2020). It is important not only to serve individual student needs but to notice and take action, as this counselor did, against the racialized segregation or *academic apartheid* that occurs in many math courses. By pointing out that the demographics of the honors class did not match the school population, and that four Black girls leaving the honors class would be problematic, the counselor made the girls' advancement a priority.

Going deep with the mathematics

Mr. Royal worked with his colleagues to better meet the different learning needs of students by critically reflecting on the purpose of the pacing guide and learning goals of the class. Slowing down the pacing allowed more time for students to make deeper connections between geometric properties, theorems, and representations. In addition, Mr. Royal also decided to introduce an interactive math journal as a strategy to support deeper engagement with mathematical ideas. The use of an interactive math journal also helped build on Elizabeth's note-taking strengths and created a supportive structure for student-teacher dialogue that invited questions and constructive feedback.

Conclusion

Sydney and Elizabeth are examples of smart, capable Black girls whose high school mathematics education has been affected by hierarchical structures like curricular tracking and the mindsets that often accompany them. It should be noted that our goal was to elucidate varied Black girls' experiences through two illustrative cases. We purposely do not want to render Black girls' experiences in mathematics as a monolith or engage in what Chimamanda Adichie describes as the danger of a single story (2009).

Elizabeth's story sheds light on how parents and caregivers play a critical role in affirming math identity and supporting learning. We see a Black mother, Mrs. Mack, exercising her agency within the school setting to advocate for her daughter and the other Black girls in the honors geometry class. Although her parental involvement may not have aligned with those of other parents (e.g., attending back-to-school night), it was clear that her mother wanted Elizabeth to develop a positive math identity and that she understood and valued the importance of her daughter staying in her math class.

In both stories, the teachers realized that instructional changes were necessary to maintain access for these Black girls. Ms. Linton and Mr. Royal took the time to reflect on their mindsets and pedagogical decisions. More importantly, they reached out to Sydney and Elizabeth about their classroom experiences and acknowledged that the girls' concerns were valid. The teachers used that information to help inform their curriculum and instructional practices moving forward. In addition, both teachers positioned the girls as strong mathematical thinkers by incorporating their strengths into their instructional changes. Doing this supports the development of positive mathematics identities and agency by Sydney and Elizabeth and also broadens the mindsets of their peers regarding Black girls' mathematics competencies.

The stories also demonstrate how colleagues can be supportive voices to nurture critical reflection and advance equity. For example, Mr. Alston encouraged Ms. Linton to get to know Sydney better and see what the class was like from her perspective. Elizabeth's school counselor stepped back and looked at the larger retention patterns of the school's advanced math courses and took action. She engaged Mr. Royal to critically reflect on the circumstances leading to Elizabeth's wanting to drop the course. She framed the issues as a problem with the system rather than a problem with Elizabeth and her friends.

The discussion in this chapter raises critical considerations about access and advancement in mathematics for Black girls. The stories highlight the harm that Black girls can experience in our current system. The strategies that we share here illuminate opportunities for Black girls to advance in mathematics education as it currently exists (i.e., honor classes). We invite you to think about what a system would be like where Black girls could thrive rather than merely survive. Within the current system, we cannot help but think of the other Black girls not in honors geometry whose brilliance is being overlooked. If mathematics classrooms are going to be places where all Black girls can demonstrate their brilliance, teachers and other school officials will need to believe in the brilliance of every Black girl. Teachers and school officials must understand how Black girls negotiate race, gender, and class and how our beliefs about race, gender, and class affect our perceptions of Black girls. We must utilize equity-based strategies that position them as strong mathematical thinkers. We must resist patterns and practices in our educational system that perpetuate deficit perspectives of Black girls and their families.

DISCUSSION QUESTIONS

1. What do you know or what can you learn about Black girls' experiences in mathematics classrooms where you work?

2. What actions can you take to affirm Black girls' positive mathematics identities?

3. What types of patterns and practices affect Black girls' enrollment and participation in advanced math courses in your education context?

4. How has your school or district partnered with Black families in service to math learning and teaching at the secondary level?

5. How can we rethink school policies such as curriculum pacing to better support Black girls' math learning in the classroom?

Chapter 9

Mathematics Assessment within Equity-Based Practices

Assessment is a multifaceted *process*.

> Mathematical competence is complex and multidimensional.... What a student knows and is able to do does not simply add up to a larger and larger amount as days go by. Progress is not a unilateral leap forward. So I would need to teach my students to see competence as multifaceted and complex. (Lampert 2001, p. 330)

The aim of the process of assessment is "gathering evidence about a student's knowledge of, ability to use, and disposition toward mathematics and making inferences from that evidence for a variety of purposes" (National Council of Teachers of Mathematics 1995, p. 3). Because mathematical competence is complex and multidimensional, tools to assess mathematical competence must be equally multifaceted to capture a child's mathematical growth and progress accurately. However, facilitating teachers' understanding of students' mathematical learning is not the only essential purpose of assessment. Communication of their learning progress to various stakeholders, especially the students themselves and their families is equally important.

Because no two students, teachers, classroom communities, or schools are alike, we do not attempt to offer a prescription for developing richer assessment tasks and practices. Instead, we provide common assessment scenarios that new and experienced teachers will recognize, including unit assessments, timed fluency tests, and problem-solving tasks, and we offer ways to rethink *routine assessment practices*. We agree that a fundamental principle of assessment is that it must first promote student learning (Black et al. 2004; Miller-Jones and Greer 2009). We extend this fundamental principle of assessment to include affirming and strengthening positive mathematics identities among students.

This chapter discusses specific ways that teachers can rethink their routine assessment practices in light of the five equity-based practices described in this book. First, we illustrate ways to provide *meaningful feedback* to students about their learning. We believe that the types of feedback that students receive can have a tremendous impact on the progress of their mathematics learning and their conceptions of themselves as mathematics learners. Next, we discuss some of the explicit ways in which teachers can recognize and tap into various forms of background knowledge and resources that students bring to assessment tasks. Consideration of these resources can serve as learning opportunities for teachers to enhance their own understanding of what students know and can do. We conclude the chapter with a discussion about rethinking assessment through revision as an equity strategy. Giving students the opportunity to revise their work based on meaningful feedback is an important part to advancing learning. We provide examples of revision policies and practices to support students to show what they know. We encourage teachers to reflect deeply on the strategies discussed in this chapter and then take action to enhance their equity-based assessment practices to improve the mathematics learning and identity development of their students.

Rethinking Assessment by Providing Meaningful Feedback

Think about the ways that students' progress in mathematics is conveyed to them. What do students see on chapter tests or homework that come back to them? A score? A percentage? Smiley faces? Check marks? Research has documented that feedback from assessments about ability and competence often has a negative impact on students, especially those from historically marginalized groups (Black et al. 2004). In particular, feedback tends to accentuate what students do not know and cannot do, thus leading them to believe that they are "not smart," lack ability, or cannot learn.

The types of feedback that students receive on assessments can contribute to their developing identities as mathematics learners. Feedback has the power to influence whether children see themselves as mathematically proficient. We encourage more holistic, positive, and specific approaches that give meaningful feedback to students with different levels of understanding. These approaches include focusing students' attention on making sense of mathematics, affirming evidence of mathematical progress (such as innovative strategies and partially correct answers and procedures), and providing students with opportunities to grow mathematically without sacrificing their mathematical confidence.

To model meaningful feedback, we showcase examples of student work along with feedback that demonstrates a range of mathematical understanding (full, partial, emergent). Our purpose here is to provide specific ways in which teachers can give students meaningful feedback that offers strategic guidance for making further mathematical progress. We believe that all students, including those identified as "high achievers," need specific feedback to affirm what they do know and can do, as well as feedback that will help orient and move their mathematical thinking forward.

Performance-based assessment task: "Floors 4 U"

We present three examples of work by sixth-grade students on a performance-based assessment task called "Floors 4 U" (Mathematics Assessment Resource Service 2010). The task assesses knowledge of area and perimeter. The real-world context of the problem is designing and measuring carpet, and students are asked to determine the amount of red carpet needed for a triangular design in a square carpet that is 8 yards by 8 yards. The student work in figure 9.1 shows the picture provided of the carpet, with the specific dimensions noted.

The figure shows three examples of student work illustrating full (student 1), partial (student 2), and emergent (student 3) understanding, as measured by the evaluation rubric. The figure shows written models of meaningful feedback and how that feedback can be linked to the equity-based practices stressed in this book, particularly *going deep with mathematics* and *affirming mathematics learners' identities*.

The aim of the feedback is to give strategic guidance to students about which aspects of their mathematical explanations are praiseworthy and which need further development (Lampert 2001). This is the core of equity-based assessment practice and can be carried out with respect to a student's level of current understanding. The feedback consists of short sentences designed to draw students' attention to specific aspects of their solutions. Orienting students' attention to productive and problematic understandings positions teachers to offer explicit guidance about what mathematical ideas to revisit and think about more deeply. This guidance is often impossible to provide with check marks or smiley faces. Further, incorporating opportunities to revise work based on meaningful feedback enables students to move forward in their math understandings in productive ways that can keep their positive mathematics learner identities intact.

We recognize that providing this type of feedback can be time-intensive initially. However, we encourage teachers to find ways to incorporate more specific feedback as an ongoing pedagogical practice. One possibility is to select a specific test or one or two problems on a test for more detailed feedback. Collaboratively creating feedback questions is also an excellent professional development activity for grade-level colleagues. Moreover, we recommend engaging students in explicit discussion—in one-on-one, small-group, or whole-class settings—after offering more detailed feedback that enables students to make sense of their progress on an assessment. Next, we explore additional ways to give meaningful feedback in routine assessment contexts, such as curriculum unit tests and fluency tests. We invite you to rethink the ways in which you communicate progress to students in relation to the models presented and the five equity-based practices.

Unit tests

Figures 9.2 and 9.3 (see pages 113 and 114) show examples of feedback on two students' work in response to a second-grade assessment item for an Everyday Mathematics lesson focused on number and operation (University of Chicago School Mathematics Project 2007). The item uses a domino representation of two sets of dots in assessing students' knowledge of fact families. Students A and B were in the same school but had different teachers. Both students were given an N (not adequate progress) for their work on this assessment item. Figure 9.2 presents student A's work and her teacher's feedback.

Student #1: Jed (full understanding)	Meaningful feedback	Link to equity-based strategies
2. The leisure center wants the carpet for this square floor to be in two different colors like the diagram. The floor is 8 yards long and 8 yards wide. How much red carpet will be needed? _32_ square yards Explain how you figured this out. 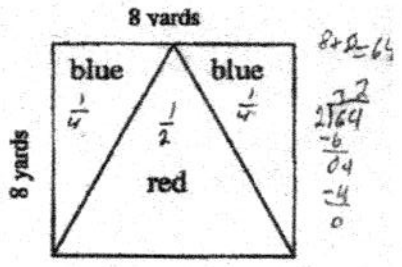 First I multiplied 8×8 which = 64, then I knew red is ½ and the 2 blues are ¼ so add them together and you get ½ so then I divided 64 by 2 and got 32.	Clear use of your fraction knowledge to explain how triangles are related to the whole and why you divided the area of the square by two. Calculations are correct. How did you know that the red triangle equaled ½ of the whole? What do you know about the relationship between triangles and squares that makes you think the red triangle is ½ and the blue triangles are ¼ of the total area?	• *Going deep with mathematics* • *Affirming math learner identity* Here the answer is correct, multiplication and division calculations are correct, and Jed shows clear fraction understanding of part-whole relationships. The feedback goes beyond phrases like "good job" and directly addresses strengths of this explanation. This is done to affirm his identity as a math learner. To deepen Jed's mathematics understanding, the questions focus his attention on examining underlying mathematical structures of the problem by asking him to make more explicit how he determined the fractional amounts and how those fractions connect to the relationship between squares and triangles.

Fig. 9.1. Examples of meaningful feedback on three students' work on a performance-based assessment task

Student #2: Desiree (partial understanding)	Meaningful feedback	Link to equity-based strategies
2. The leisure center wants the carpet for this square floor to be in two different colors like the diagram. The floor is 8 yards long and 8 yards wide. How much red carpet will be needed? ___32___ square yards. Explain how you figured this out. *[handwritten diagram: square divided by diagonal lines, labeled 8 yards on top and 8 yards on side, with "blue", "blue", "red", "blue" regions, and "= 64"]* I figured out by because if you divide 2 into 64 you will get 32 square yards. I got two because 2 is a 1/2	Revision suggested: Good use of your knowledge about partitioning as a method to find the area of the square. It is correct that 2 divided into 64 equals 32 square yards. However, you need a clearer explanation that connects how you solved the problem with the picture. For instance, how does the partitioning help you figure out the area of the triangle? Where does 64 come from? Why did you divide 64 by 2? Explain why "2 is a $\frac{1}{2}$." How does 2 or $\frac{1}{2}$ relate to the area of the triangle?	• *Going deep with mathematics* • *Affirming math learner identity* • *Drawing on multiple knowledge resources* Here the solution is correct. However the explanation needs clarity. The division statement is correct that 2 divided into 64 equals 32. The student also used a common method of partitioning a figure into unit squares. However, the student claims that "2 is a $\frac{1}{2}$," which is not true. There is also no direct link to the area of the red triangle. The feedback suggests revision followed by noting the accurate method used and the correct division statement. Acknowledging the strengths of the work helps to maintain math confidence. However, it is equally important to help Desiree refine her mathematical argument to justify her correct answer. The questions point Desiree to rethink the connection between her explanation and the picture, her chosen method of partitioning, as well as clarify her claim that "2 is a $\frac{1}{2}$." To make forward progress, the final question orients her to consider how 2 or $\frac{1}{2}$ might be related to the area of the triangle, thus pushing her to deepen her understanding of area and the relationship between the triangles and square.

Fig. 9.1. Examples of meaningful feedback on three students' work on a performance-based assessment task, *continued*

Student #3: José (emergent understanding)	Meaningful feedback	Link to equity-based strategies
2. The leisure center wants the carpet for this square floor to be in two different colors like the diagram. 8 yards 8 yards The floor is 8 yards long and 8 yards wide. How much red carpet will be needed? 39 square yards Explain how you figured this out. I found the area first and made squares on the triangle and counted them, and it was 39	Revision needed: Good partitioning model for understanding the area of the square. Sixty-four square yards is the area of the square. However, your counting strategy of how you got 39 in the red triangle is unclear. How are the red and blue triangles related to the square in this picture? What part of the whole square do they represent? How could you use what you know about the area of a square to help you find the area of the red triangle?	• *Going deep with mathematics* • *Affirming math learner identity* • *Drawing on multiple knowledge sources* Here the answer is incorrect and the explanation is limited. The feedback starts with a statement that revision is needed followed by a positive statement that affirms clear use of an appropriate mathematics strategy to find area of a square—partitioning the figure and counting the unit squares. However, this partitioning strategy can be limiting with more complex figures, particularly with triangles. This resulted in a miscount of unit squares. It is important to acknowledge the strategy and point out how it is problematic. To deepen Jose's mathematics understanding, the comments focus his attention on the relationship between the triangles and square and part-whole relationships. The feedback is designed to connect to other mathematical knowledge about part-whole relationships and areas of specific shapes José may have but did not use in this problem. This feedback can help guide his revision.

Fig. 9.1. Examples of meaningful feedback on three students' work on a performance-based assessment task, *continued*

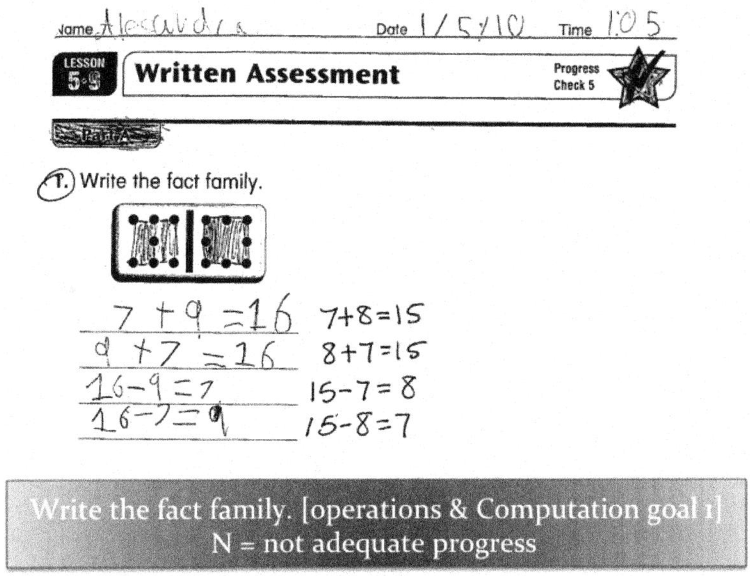

Fig. 9.2. Sample work from student A and teacher feedback

As the figure 9.2 shows, the teacher's feedback on this second grader's work is very straightforward. The correct four-fact family connected with the domino representation is written next to the student's four-fact family number sentences. However, the feedback makes no explicit connection between the student's solution and the teacher's correction other than their close proximity. One issue to consider is whether student A does show knowledge of fact families. She provides four mathematically correct number sentences, consisting of two addition sentences and two subtraction sentences, which include the numbers 7, 9, and 16. Clearly, she did not accurately count the number of dots on the right-hand side of the domino. If we think about ways to draw this student's attention to the mathematics to make sense of it, what question or comment (written or oral) could we offer to extend her understanding about fact families?

One way might be to acknowledge student A's demonstration of understanding of a fact family. Using the approach illustrated in figure 9.1, the teacher could acknowledge the student's counting error but write something like "Good use of fact family addition and subtraction equations for the numbers 7, 9, and 16." Then the teacher could follow up by drawing an arrow connecting the fact family written by the student and the domino with a question such as "Do these match?" In this way, the teacher could recognize this student's growing understanding of fact families while drawing her attention to the discrepancy between the pictorial representation in the domino and the number sentences depicting that representation.

Consider the second example involving a student's work on the same problem with feedback from a different teacher. Student B's work appears along with teacher feedback in figure 9.3.

In this case, the teacher conveys the feedback through check marks on two of the four number sentences. The check marks are attached to two subtraction number sentences. Although one subtraction sentence (8 − 7 = 1) is mathematically correct, it is

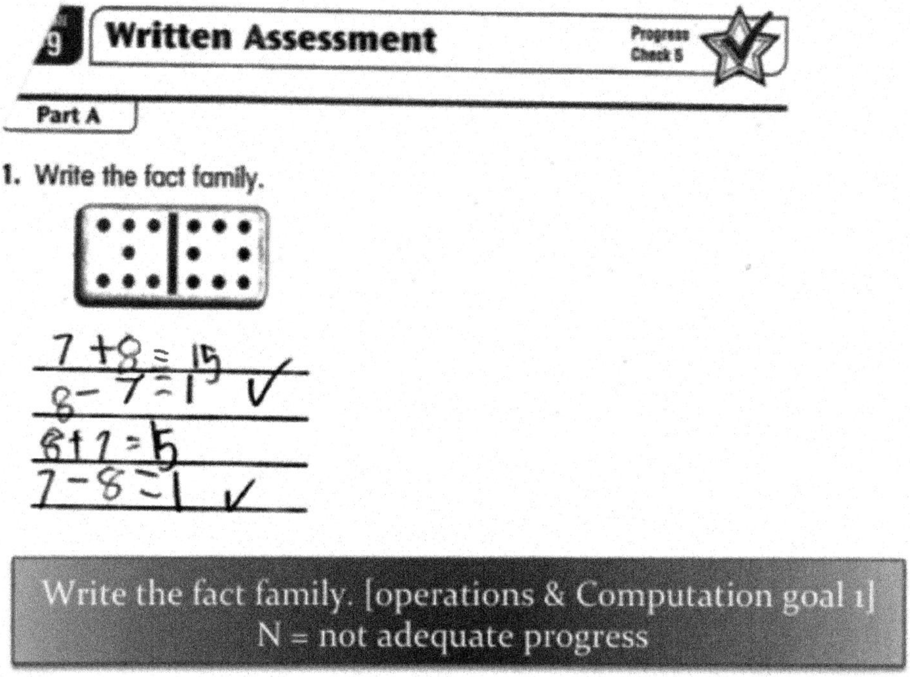

Fig. 9.3. Sample work from student B and teacher feedback

not a fact family number sentence that reflects the domino representation. The second subtraction sentence is clearly incorrect. Does this student demonstrate knowledge of fact families? How does the feedback assist the student in deepening their math understanding of number relationships by working with fact families?

If we focus on the primary aims of promoting student learning and positive mathematics identities, this kind of feedback gives us pause. Clearly, this student's understanding is still emerging. Although the addition equations are correct and match the representation, the subtraction equations tell a different story, raising questions about the student's understanding of number relationships. Meaningful feedback would require more specificity in this case to deepen this student's mathematical understanding. An important point to make with this student is that the addition equations need to "match" the representation. However, focusing on the subtraction equations, we might acknowledge the correctness of the first subtraction but point out that the subtraction equation does not match the domino picture. We could write something like, "Almost! The addition equations of the fact family match the domino. But does the subtraction equation match? Does $8 - 7 = 1$ match the picture?" Then the teacher could draw an arrow to the picture and list the specific numbers (7, 8, 15) that should be in all four equations.

These two examples of the same summative assessment item are designed to promote reflection about how feedback is often given to students, and by extension, to their parents or caregivers, who might be reviewing the test results. The goal here is to provide feedback that promotes deeper mathematical understanding, acknowledges what the child did correctly, and guides the child's attention to misunderstandings that need to be addressed. Teacher feedback plays a crucial role in the process.

Algebra 2 quiz

Providing meaningful feedback is also helpful in math courses at the high school level. Figure 9.4 shows teacher feedback on one quiz problem related to nonlinear functions. The original teacher feedback affirms student identity by noting specific strengths in the new graph while also raising awareness of an area for growth: "Great job shifting the graph. Next step, let's clearly label key points." This teacher's brief feedback demonstrates attention to specific strengths, noting "Good" as a way to signal correct answers. The teacher also asked open-ended questions to further promote student thinking such as, "How did the function shift?" This feedback provides a solid foundation for the student to understand their mathematical strengths and areas of growth. Building on the teachers' comments, we offer expanded feedback that could further specify a productive learning path for this student.

Original Feedback	Expanded Feedback
	Input-output table is a good way to generate points. Can you check your calculations and add other points to the table?
	Great job shifting the point (0,0) to (−3,−2). How did your other key point shift? What other key points could you add to your graph? How did they shift?
	Can you use your graph and domain and range statements to describe how the function shifts?

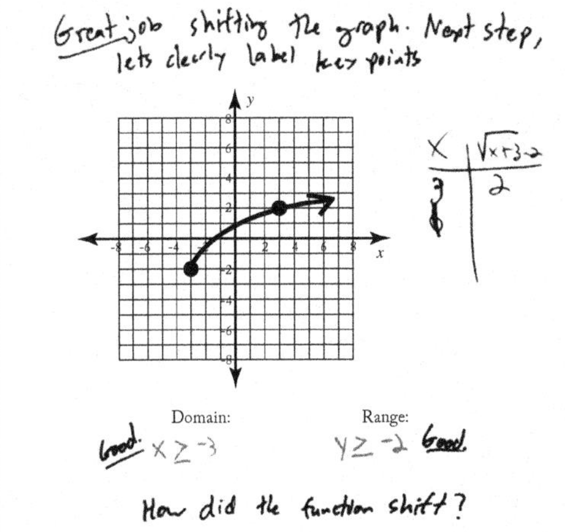

Fig. 9.4. Teacher feedback on algebra quiz problem

A key takeaway here is to provide specific feedback that invites students to make connections to their work (affirming their strengths) while asking strategic questions that support students to go deeper with the mathematics.

Fact fluency assessments

Michael Allen, a fifth-grade elementary teacher, is one of the six new teachers whose mathematics autobiographies are presented in Chapter 3. Michael describes his early mathematics experiences:

> My earliest math memories are of timed fact fluency tests in third grade. I remember that I was disappointed to not be one of the fastest in my class. However, I also remember being relieved that I was not one of the slowest either.

Although Michael is a teacher now, his earliest memories of mathematics learning focus on a widely used assessment for multiplication and division math facts—also known as "timings"—short tests with 20–60 problems of similar type that students are expected to complete within a short period of time (1–3 minutes). What is important to note about Michael's words is the relationship of these timings to his identity as a mathematics learner. Being "slow" or "fast" was determined by this kind of assessment and brought both disappointment and relief to Michael as a young student.

Because of their widespread use, fact-fluency timings and the role that they play in students' mathematical learning and mathematics identities are important to think about. Students do need to develop robust forms of procedural fluency that emphasize efficiency, accuracy, and flexibility (Russell 2000). Although these assessments may work to develop efficiency and accuracy through rote memorization, they do little to promote flexibility and understanding of procedures (National Research Council 2001a). Furthermore, as reflected in Michael's comment, speed is connected to status and "smartness" in mathematics. Being slow puts students at risk for marginalization. Thus, although the intent of such assessments is to strengthen a specific mathematical skill, the consequences for students' mathematics identities are far-reaching and need to be considered.

Figure 9.5 provides an example of a typical way in which teachers construct and score such timings. Feedback is usually given in the form of a fraction: total number of correct answers over total number of problems (in this case, $^{20}/_{26}$). Incorrect answers are noted by a check mark. The time taken to complete the task is also noted in minutes and seconds (1:52).

This form of feedback—check marks—emphasizes the mistakes made. In some cases, students might have to take this test over until they reach 100 percent accuracy before moving on to the next fact timing. Over time, if students continue to struggle with accuracy or efficiency (by not completing all the problems in a given time limit), this process can diminish their confidence as mathematics learners.

How can feedback on timings be made more meaningful? One approach is to do a close analysis of the student's work, looking for specific patterns in correct and incorrect answers. Are there specific consistencies or inconsistencies? Here we extend traditional error analysis (Ashlock 2002) to a more balanced "strength-need" analysis of the students' work to address their mathematical learning progress more effectively.

In the work displayed in figure 9.5, the student is consistently correct in her use of the following four-facts:

$$4 \times 0, \quad 4 \times 1, \quad 4 \times 2, \quad 4 \times 3, \quad 4 \times 4, \quad 4 \times 5, \quad 4 \times 6, \quad 4 \times 10, \quad 4 \times 11, \quad 4 \times 12$$

Furthermore, the student appears to understand the commutative property of multiplication (that is, $a \times b = b \times a$) in all the four-facts with correct and incorrect solutions. Emphasizing the four-facts that the student has learned with efficiency and accuracy will help the student—and the parents and caregivers who assist the student at home—to be more strategic about strengthening fluency with all the four-facts. Figure 9.5 also reveals a group of four-facts that still needs work:

$$4 \times 7, \quad 4 \times 8, \quad \text{and} \quad 4 \times 9$$

Name __Alesandra__

fours

10×4 __40__   1×4 __4__

0×4 __0__   3×4 __12__

4×1 __4__   4×6 __24__

4×11 __44__   ✓9×4 __38__

4×2 __8__   ✓7×4 __30__

4×3 __12__   4×4 __16__

6×4 __24__   11×4 __44__

4×4 __16__   4×10 __40__

12×4 __48__   1×4 __4__

5×4 __20__   ✓4×8 __34__

✓4×7 __30__   4×0 __0__

✓8×4 __34__   2×4 __8__

✓4×9 __38__   12×4 __48__

Score: __20__ /26

Date: __3—7—12__

__1:52__

Fig. 9.5. Fourth-grade multiplication fluency test

Although the feedback provided by check marks is immediate, it is not always meaningful, particularly for indicating how students can strengthen fluency with multiplication facts. One way to give such feedback is to focus student attention on the meanings, interpretations, and models of multiplication and their relationships to known facts, thus bridging from the facts that students do know to those that they have yet to learn. This approach emphasizes flexibility, a third component of procedural fluency, along with efficiency and accuracy (Russell 2000). Written feedback that draws student attention in this way on timed tests on four-facts could include statements such as the following:

- Almost there! You've got most of your four-facts down. Let's work on 4 × 7, 4 × 8, and 4 × 9. Let's bridge and use what you already know! How can the facts 4 × 1 = 4 and 4 × 6 = 24 help you remember 4 × 7? How can the facts 4 × 1 = 4 and 4 × 10 = 40 help you remember 4 × 9?

- Close! We just have to work on the middle facts (4 × 7, 4 × 8, 4 × 9). What is the relationship between 4 × 4 and 4 × 8? Use our doubling strategy. You can do this!

The key here is to provide meaningful feedback to deepen students' understandings and to help them connect their previous knowledge with what they still need to learn. Focusing the feedback on mathematical properties such as the distributive property and strategies like doubling can help students make these connections. Further, this type of written feedback is positive, encouraging, and strategic. It affirms the multiplication facts that students know, suggests ways to bridge to facts that they have yet to learn, and affirms their identities as mathematical learners moving forward.

Rethinking Assessment to Engage Multiple Knowledge Resources

During mathematics lessons, teachers can use multiple forms of assessment to recognize and engage the multiple forms of background knowledge and resources that students bring to mathematical problem-solving situations. They can use these to develop more holistic understandings of what their students know and can do. One of the equity-based practices specifically focuses on *drawing on multiple resources of knowledge*, including mathematical, cultural, linguistic, family, and community resources. These resources are grounded in the social realities and experiences of students and can contribute to their learning and the development of positive mathematics identities. We suggest that teachers can learn more about how students are making sense of mathematical ideas by paying attention to these resources as they are expressed and communicated by students in written and oral presentations of their work. We discuss two examples that can help teachers rethink their assessment practices and maximize their opportunities to learn more about the resources that students bring to bear on math problems.

Valentine exchange activity

During a fourth-grade math lesson, the teacher, Mrs. Olivas, carefully posed a series of questions that asked students to determine the number of valentines exchanged among the twenty-four students in the class at the recent Valentine's Day party. Each student had exchanged one valentine card with every other student. To launch the lesson, Mrs. Olivas posed simple scenarios involving exchanges among two people, three people, and four people, asking her fourth graders to predict a solution and then model the valentine card exchanges physically to verify the predictions. Her launch of the lesson was limited by her use of only one valentine when modeling the exchange. However, her students developed and clarified their understanding by modeling the exchange that had occurred previously at the party. (The teacher's and students' real names are used in this example; for edited footage of the "Valentine Exchange" lesson, see Annenburg Media [1995].)

Figure 9.6 shows a re-creation of a drawing made on the board by Armando, a Spanish-dominant fourth grader in Mrs. Olivas's class. This drawing is a pictorial representation of Armando's solution to the question that Mrs. Olivas posed orally: "If there were four people that exchanged valentines, how many valentines would be exchanged?"

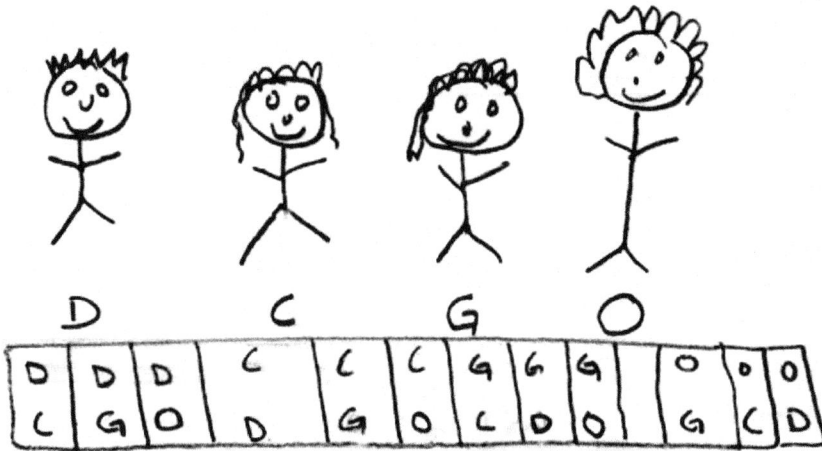

Fig. 9.6. A re-creation of Armando's blackboard drawing

While Armando explained his solution in Spanish, he pointed to the different combinations represented in the rectangles and counted in Spanish: "uno, dos, tres, cuatro, cinco, seis, siete, ocho, nueve, diez, once, doce," thus demonstrating that there would be twelve exchanges among four people. The four people in the picture represented three classmates, Denry (D), Chelsea (C), and Gina (G), and the teacher, Mrs. Olivas (O), who had just physically modeled the exchange of valentines.

Armando's representation provided Mrs. Olivas with a clearer understanding of his mathematical thinking, including the resources that he brought to bear in this explanation. His picture modeled different mathematical features of the situation—the people involved and the different permutations of card exchanges. He accounted for each person exchanging a card with another person by recording the initials of each person involved in an exchange in a rectangle below the pictured person. In this lesson, English was the primary language of instruction. But Armando's explanation combined his visual representation with gestures, such as pointing to specific rectangles to convey his mathematical understanding of this problem. By inviting Armando to explain his solution in his first language, as well as through pictures and gestures, Mrs. Olivas enabled Armando to make a valuable contribution to the mathematical discussion that benefited his English-only and bilingual peers (Aguirre et al. 2012; Moschkovich 1999). A multilingual student, Armando was not relegated to the margins or ignored during this lesson. He participated as a legitimate classroom member, offered an alternative reasoning strategy and a representation to solve the problem, and received affirmation of his mathematics identity along with his linguistic identity (Zavala and Aguirre 2023).

Armando's solution provided solid evidence of a reasoning strategy that could be leveraged to further develop skills in pattern identification and generalization to determine how many valentines would be exchanged in larger groups, such as the class of twenty-four students, the entire fourth grade, or any number of people. In the future, Mrs. Olivas could offer further support for Armando's mathematical understanding by helping him recognize the limitation of using pictures as an efficient representation for valentine card exchange combinations with larger groups of people. At the same time, she could use his representation as a reference to help him notice a pattern among the cards exchanged: each person gives a card to someone else except for themselves (that is, the picture has no rectangles with identical initials, such as D, D). Helping Armando see the strengths and limitations of his solution is an equity-based assessment strategy that can deepen his mathematical understanding.

Bus Pass problem

Figure 9.7 presents a best buy problem that was given to students in a predominantly African American urban middle school (Tate 1994a). Many students reasoned about and solved the problem in a way that test designers did not anticipate.

Bus Pass Problem

It costs $1.50 each way to ride the bus between home and work. A weekly pass is $16. Which is the better deal, paying the daily fare or buying the weekly pass?

Fig. 9.7. The Bus Pass problem

The assumption of the district's test designers was that students who solved the problem correctly would choose to pay the daily fare. When many students answered the problem "incorrectly," teachers discussed reasons for this outcome with their students. The teachers discovered that the assessment item was based on a particular set of assumptions that privileged students accustomed to thinking in relation to an eight-hour weekday work schedule and people having only one job. However, many African American students solved the mathematical problem under a different set of assumptions based on their daily lives and social realities.

When the teachers asked some African American students to explain their choice of the weekly bus pass, the students indicated that the weekly pass was the better option because it could be used in a variety of ways that were work and non-work related. It could also be used on weekends or by other family members. The African American students made accurate mathematical computations in the context of the specific scenario; however, their ultimate selection of the weekly bus pass over the individual daily fare as the better deal was based on different assumptions, grounded in their home and community experiences.

Although many arguments can be made about why the Bus Pass problem is or is not a good assessment question, we focus here on the *learning opportunities* that such a question provides for teachers, enabling them to reflect on the equity-based practices promoted in this book. The important point in this context is that mathematical understanding is not disconnected from students' social realities. In fact, students with different backgrounds and experiences draw on these to make sense of ideas and problem-solving situations. The Bus Pass problem uncovered the different mathematical assumptions made by the students and the test designers, grounded, perhaps, in different social realities. It is important to recognize and engage these kinds of assumptions in the classroom to get a better understanding of what students know and can do.

For example, after closely examining student work and the responses that differed from the solution presumed to be correct, teachers might have asked the following question as a way to probe the mathematics more deeply with their students: "What mathematics skills, concepts, and thinking strategies did you actually use to derive your solutions to the problem?" In the Bus Pass scenario, students would have had to demonstrate computational accuracy and fluency in determining the total amount paid based on individual fares. But because many of the students reasoned about the possibility of multiple uses, the problem also allowed students to demonstrate skills in strategic reasoning and problem solving.

The assessment item also showcases how students draw on their everyday and cultural experiences to make sense of mathematics problems presented in school. This kind of assessment task can open up opportunities to leverage multiple mathematical competencies and make transparent the mathematical and background resources that students draw on to solve the problem. Moreover, given that students are likely to be personally invested in their solutions, they may be more likely to share and justify their thinking out loud and give meaning to the mathematics they used to generate the solutions. In doing so, students recognize that their mathematical contributions are valued, and thus, the spaces in which students are marginalized may shrink.

If students struggle to generate meaningful solutions on their own, teachers can ask them to build on one another's partial solutions. This activity could help to create classroom environments where the teacher and students can leverage one another's multiple competencies. When students generate various solutions to a problem, as in the case of the Bus Pass problem, teachers can use this situation as an opportunity not only to confirm the possibility of multiple solutions but also to explore alternative strategies, assumptions, and solutions generated by students. By acknowledging multiple student solutions and thinking processes, teachers can positively affirm students' mathematical identities, showing that equally valid solutions can be generated by different students, thus affirming all students as powerful "doers of mathematics" and changing the nature of mathematics classrooms from places where "smart" and "not smart" student mathematics identities emerge.

The "Valentine Exchange" lesson and the Bus Pass problem highlight ways teachers can rethink assessment to take advantage of the mathematical resources that students bring to mathematical problem solving. It is important to remain alert for these mathematical resources when they emerge in a lesson and recognize and engage them as essential sources of information for understanding what students know and can do. Paying close attention to how social realities inform mathematical assumptions and shape learning will benefit instructional decision making that supports students' mathematical progress and strengthens their mathematical identities.

Rethinking Assessment and Grading Policies to Support Student Learning

Most public school districts give standardized tests at least once a year. Due in part to the Common Core State Standards, the format and content of many tests have changed from multiple-choice items to a variety of item types such as enhanced multiple-choice, evidence-based selected responses, short answers, and extended-response items. Although some items may allow for more than thirty to sixty seconds per question, for the most part, they are still summative and in on-demand format, where students only have one chance to demonstrate their knowledge, and their score is based on that day's measure. One can argue that this format is necessary for large-scale state-mandated testing, but what other rationales exist for its use at the classroom level?

Educational psychologists categorize education's traditional testing methods as *static*, meaning a testing regime that captures a narrow range of students' knowledge based on their ability to demonstrate it at a specific point in time and without the contextual factors that support that knowledge (Dumas, McNeish, and Greene 2020). Some schools are trying to infuse a cycle of testing that centers on *revision* to expand students' learning opportunities and focuses on growth. These revision-focused strategies demonstrate at least three equity-based practices: deepening students' mathematics understanding by responding to the meaningful feedback offered on assessments, affirming a student's mathematics identity by promoting persistence, and challenging spaces of marginality by distributing mathematics authority. The following examples from elementary and high school settings offer ways to rethink on-demand testing by emphasizing the impact of incorporating test retaking and grading policies.

Fraction-unit test in fourth grade

Ms. Harper, a white fourth-grade teacher, works at an urban elementary school in the southwest United States. She just completed grading her recent mathematics unit test. She is satisfied with the levels of understanding demonstrated by most of the students. She makes notes in her planning journal about the types of errors made so she can incorporate them into her reengagement lesson. However, she is concerned about Kali, who did not finish the test. When Ms. Harper asked Kali why she did not finish the test, Kali's eyes filled with tears, and she shrugged her shoulders. Kali is a nine-year-old African American girl who tends to be rather quiet in class. She has good study habits, does all her homework, and gets along with all her classmates. Back-to-school night was coming up soon, but Ms. Harper thought it would be a good idea to talk with Kali's parents as soon as possible. Her mother is very active at the school and is one of the classroom parents.

Ms. Harper begins the meeting by telling Kali's mom, "I wanted to talk with you to learn more about Kali's previous math experiences." Ms. Harper knew from the class autobiography project that Kali's least favorite subject was math, and her favorite was writing. Kali's mom explained,

---

Kali had a difficult first-grade year in math. Her teacher emphasized speed tests and used competitive games as motivation to study. Kali did not respond well at all to those strategies. When put on the spot, she froze. Kali likes to please people, especially her teachers. If she doesn't think she can do well, she gets scared and shuts down. We've worked hard the last few years on increasing her math confidence but sometimes her anxiety really kicks in on tests. I know she understands fractions. She helps me with the baking for our church and does all the measuring when we have to double or triple our batches.

Ms. Harper noted in her journal to rewrite some of the problems as recipes, building on Kali's familiarity with fractions in cooking. They brought Kali into the meeting and asked her about the incomplete problems. Kali said, "There were so many problems. I looked at all of them and knew I couldn't finish in time." Ms. Harper asked her if she understood how to solve the problems, and Kali said, "Some of them." Ms. Harper suggested that she and Kali go over the problems a couple at a time together. Ms. Harper also emphasized that Kali could have as much time as she needed to complete the problems. Ms. Harper told Kali she would have a chance to retake the test when she felt ready. She also showed her a strategy of folding the paper in half so she could see only a few problems at a time. After leaving the meeting, Kali and her mom felt hopeful about this year's math experience.

Ms. Harper decided that a single test score was insufficient to determine Kali's understanding or assume her lack of understanding of the material. More information was needed. Ms. Harper saw Kali's mistakes as an opportunity to learn more about her math identity and how this might affect her math performance on summative assessments. As it turned out, Kali's mother had a wealth of information that proved to be very helpful in reaching out to Kali. Knowing that Kali cooks with her mother and seems to have a conceptual understanding of fractions will enable Ms. Harper to build bridges from Kali's lived experiences with fractions to the type of fraction problems in this unit. She could draw on multiple knowledge resources from Kali's family and community experiences. Allowing Kali to retake the exam after they have had an opportunity to delve deeper into mathematics also supports Kali's positive math identity development. Kali can begin to see her mathematics learning as a process, not an end goal. Knowing she can continue to learn will increase Kali's willingness to persevere when she faces problems that she is unsure how to solve.

"It's not over": A high school assessment policy

Mr. Peters, a principal at a college-prep high school, decided he wanted to address the inconsistency in grading practices across content areas. He was concerned when he began to analyze testing and grading practices at his school. Having studied the work of education leadership scholars such as Guskey (2014), Stiggens (2017), and Feldman (2018) on equity grading and assessment practices, Mr. Peters noted that the traditional practice of taking scores and averaging them did not adequately reflect student knowledge. Student behavior was also a component of a student's grade, which worked to the benefit of some students and the detriment of others, mainly students of color. Mr. Peters wanted to implement a new grading and exam policy (see figure 9.8). Several faculty meetings were held. The staff looked at data on student grades across and within departments. Teams from various departments attended several professional development sessions focused on assessment and shared information at their department meetings. There were many tense conversations about assessment changes, but a new policy was implemented two years later. The school decided to separate behavior and study habits from course grades. Students would receive two types of grades: Habits of Scholarship Grade and a Course Grade. They also adopted a new testing policy that would apply to all content areas.

Retesting Policy

Any major summative exam—except for the final exam—can be retaken once for *full replacement* of the original grade under the following conditions:

1. There will be one opportunity to retest per major test/assessment.

2. Students have a right to retest provided they show evidence of observable, consistent, assiduous habits of scholarship.

3. Students who exhibit proficiency (i.e., a B or better) may be advised not to retake the assessment. If a student has exhibited proficiency on the exam, the retake may have a negligible effect on the overall learning (and grade) in the course, and subsequent course assessments provide opportunities to show mastery.

4. Students must take the retest opportunity during A-Block within *five school days* of the return of the original exam. *(A-Block is the last period of each day where students can get help from teachers or counselors, meet with advisors, retake exams, etc. No extracurricular activities can begin until after A-Block.)*

5. Teachers reserve the right to offer students a retake of part or all of the assessment, the results of which *replace* the portion of the exam that was retested, *even if the student performs less well on the retake.*

Fig. 9.8. High school retesting policy

This was a big change for students as well. One student interviewed for the school paper about the school's new schedule and assessment policy stated:

I like A-Block and our new assessment policy. A-Block is extremely helpful when you want to get a head start on your work that's freshly in your brain or to get questions answered. I often overthink things when taking math tests, and the last test I didn't do so well. Part of the mandatory process for retesting in math is to go over the test with your teacher during an A-Block and redo problems that you got wrong. Finishing the corrected problems gives you a few extra points for your Habits of Scholarship grade. Sometimes, my teacher sets up study groups during A-Block. Reviewing the material with my classmates is helpful too. We go over problems together that each of us don't understand and help others on the topics we feel confident about. This part is really helpful. Teaching someone else, talking aloud, and writing helps me remember the information even more. I've taken two retakes this year. Each time I was able to do better than the first time. I don't feel as bad if I don't do well on a test. It's like I have another chance. It's not over.

From this student's standpoint, retakes are opportunities for students to demonstrate their best work. The revised assessment policy is more holistic. Offering more time and collaboration helps students to revisit and revise their understanding.

Three equity-based practices are embedded in this policy: going deep with the mathematics, affirming mathematics learners' identities, and challenging spaces of marginality. When the mathematics department made adaptations to the policy requiring teachers to meet with the students after the exam, they provided the opportunity for teachers to elicit information about student thinking and responses. Inquiring about their thought processes preceding and during the exam enables teachers to ask questions that uncover mathematical strengths and confusions that were not evident in the exam. It also provides the teacher with information to target areas of growth and make connections between what the student understood and the new knowledge, enabling students to deepen their mathematics understanding. Additionally, students with a less positive mathematics identity often are reluctant to seek assistance. They may be embarrassed to tell their peers why they must miss a sports event or club activity. Creating time for students to seek assistance and integrating it into the schedule (A-Block) demonstrates its priority in learning and, most of all, normalizes it for students, encouraging broad-based participation and thus challenging students' spaces of marginality. The student feels more comfortable talking with the teacher about mistakes instead of asking in class. Also, being able to work with peers to ask for and provide help both affirms a mathematics learner identity and distributes math authority among the group. Separating habits of scholarship from a course grade forces attention on mathematical thinking and communication, not on perceived behavior factors that are often the focus and can cloud a student's brilliance from being seen.

Revisiting concerns about retakes

There are instructional concerns raised with retesting as well as with providing feedback. Some teachers, especially in middle school and high school, might think, "There are too many students to allow retesting. What's to keep students from slacking off and not studying just to buy more time? How is retesting fair to other students who were able to show mastery on the first test? How is that preparing them for college?" Each of these questions brings us back to fundamental questions about teaching, learning, and assessments. Why are we giving a test? Does everyone need to learn at the same pace? If summative tests are a measure of students' content knowledge, and if we truly believe all students are capable of learning the material, then how fast one achieves the goal may be overstated with respect to its importance. Additionally, while it is true that most college math departments do not have retesting policies, there are college math teachers who have adopted methods that seek to decrease test anxiety and support student interaction and community. Such a case is seen in a Calculus II course taught at a large public research institution, where the teacher sought to rehumanize the structure of his exams by dividing them into solo and group sections. The group portion contained the most difficult items based on the previous year's data. Students had individual time to work on problems and then group time to collaborate and reason collectively (MacArthur 2019).

Adopting a "do no harm" perspective requires us to provide as much latitude as we can within the testing regime if we cannot escape it. The goal is learning, and tests are powerful tools that, if not leveraged in a humanizing way, can stifle, and ultimately extinguish, students' belief in their ability to be and see themselves as mathematical thinkers.

Rethinking the grading process to sustain positive math identities of students

In Chapter 4 we discussed the ways standards and standard-based practices are often used to perpetuate issues of antiblackness, xenophobia, and white supremacy.

They position teachers to view students as *those who can* and *those who cannot*, and this mindset can undermine a teacher's equitable instructional goals. One of the most pervasive ways this can occur is through the relationship between assessments and grades.

Assessment should be a tool to inform teachers how well a student understands what is taught and for students to synthesize what they know about a topic. Unfortunately, in practice, assessment is often *equated* with grading, which moves its functional purpose for teachers into one where categorizing and labeling students is the objective. High-stakes state exams, like the Smarter Balanced Assessment Consortium (SBAC) given in third through eighth and eleventh grade that categorize performance as *exceeded*, *met*, *nearly met*, or *did not meet*, often determine whether a student proceeds to the next course or level. As discussed in Chapter 5, this leads to students being compared with each other, sorted into a hierarchy of ability, and placed into tracks that result in different learning opportunities and experiences (Berry, Ellis, and Hughes 2014; Boaler and Selling 2017).

Providing positive feedback to students on performance-based assessment tasks or unit tests strengthens a student's identity and deepens their mathematical knowledge. But this is not enough. Research has shown that students tend to ignore feedback when accompanied by a grade or overall judgment (Black et al. 2004; Yeager et al. 2014). Feedback should be seen as a formative process, not a summative one. A numerical mark or grade does not tell you what to do: if it is high, you're pleased but have no impetus to do better; if it is low, it reaffirms your belief that you are not able to learn the subject. Looking back at an example of meaningful feedback presented earlier in the chapter, in figure 9.1, Jed's answers were correct. Providing him with a check or a letter grade doesn't foster Jed's thinking. The feedback question, "What do you know about the relationship of triangles and squares that makes you think the red triangle is $1/2$ and the blue triangles are $1/4$ of the total area?" invites Jed to think deeply about the mathematical structures in the problem. Furthermore, if a high school student who got a C on their test wasn't allowed to retake the exam, the student's learning would have ended and the confusions might persist. Meaningful feedback and opportunities to revise one's thinking by way of flexible retake policies can have a positive impact on a student's mathematical identity.

Conclusion

Assessment is a key component of all teachers' instructional practice. It can be a vital tool aimed primarily at promoting learning and positive mathematics identities in students. Providing meaningful feedback enables students to make sense of their current and future mathematical progress. Engaging the multiple forms of knowledge and background resources that students bring to the classroom and using them to help students gain a better understanding of mathematics to extend their mathematical progress and confidence is fundamental to equity-based practice. Finding ways that encourage students to make sense of mathematics, draw on their previous knowledge, and affirm their existing competencies should be key goals for all teachers. We encourage teachers to identify and critically examine assessment and grading policies that dehumanize students by asking key questions such as, Are these assessment practices worthy of my students? Can these assessments be implemented in ways that affirm a wide range of productive mathematical identities? Do they afford opportunities for all students to gain access to identities as doers of mathematics? We want all students to be able to say, "I don't feel as bad if I don't do well on a test. It's like I have another chance. It's not over."

DISCUSSION QUESTIONS

1. How do you communicate mathematical accomplishment or progress to students?

2. Select a sample of student work. Based on some of the examples of meaningful feedback provided in this chapter, decide what kinds of comments you can write to extend the student's mathematical thinking and preserve a positive mathematics identity.

3. In what ways have you noticed students using their everyday lives and experiences to make sense of and solve math problems?

4. How do your mathematics lessons provide opportunities for you to learn more about the multiple resources that students bring to the work of solving math problems?

5. How do you feel about retakes on summative assessments? What are the benefits and challenges?

6. What other ways can teachers support students to revise their thinking?

Part 3

Rethinking Engagement with Families and Communities

Part 3 encourages teachers to reflect on the roles that parents, caregivers, and community organizations can play in deepening mathematics learning for students and reinforcing students' positive mathematics identities. Rethinking these roles may require additional effort. However, the benefits for classroom instruction, student engagement and learning, family participation, and community support are well worth the time.

Chapter 10 focuses attention on routine classroom interactions with parents and caregivers about mathematics learning. Newsletters and back-to-school night activities are modeled as ways to communicate a teacher's mathematics vision and showcase equity-based practices to help children learn mathematics. Specific communication strategies in the context of parent-teacher conferences are also modeled as ways to have a positive impact on students' identities. The chapter provides both conceptual and practical tools that can help teachers enlist parents and caregivers to offer further support for children's mathematics learning and development.

Chapter 11 describes mathematics programs that partner with parents, caregivers, and community-based organizations to support mathematics learning *outside* the classroom. The chapter highlights two nationally recognized parent education programs—FAMILY MATH and MAPPS (Math and Parent Partners project)—which provide parents and caregivers with tools and perspectives to support their child's mathematics education at home. An example also illustrates school partnerships with community and faith-based organizations to provide after-school mathematics support to students.

Chapter 12 examines making connections with families at the secondary level. The chapter takes a deep dive into various ways that teachers, schools, and districts communicate with families about mathematics policies and practices that can influence students' math identities, such as math pathway decisions. Through illustrative vignettes, educators are encouraged to reexamine their communication strategies and the well-being of family partnerships to support student learning.

We encourage a paradigm shift in thinking regarding the roles and participation of parents, caregivers, and community organizations in a student's mathematics education. In-classroom and out-of-school examples can assist teachers in developing true partnerships with parents, caregivers, and community groups to support students' mathematical, social, and knowledge empowerment.

Chapter 10

Routine Practices to Engage Parents and Caregivers in Promoting Positive Math Learning and Identity

This chapter describes ways to enhance *routine teaching practices* by involving parents and caregivers in supporting mathematics learning and positive mathematics identity development among students. Our goal in stressing equity-focused routine practices is to *enhance what you may already do*, as well as introduce a few new options that might offer you possibilities for increasing parent and caregiver engagement and strengthening their relationships with teachers. We will focus on two types of routine classroom interactions with parents and caregivers. First, we will explore common routines and events that many teachers use to introduce their mathematics practices to families—specifically, classroom newsletters and back-to-school night events. Over the school year, these are likely to offer you primary opportunities for communicating your vision of mathematics to families and keeping them updated on the mathematics in your classroom. Second, we will examine classroom interactions related to mathematics progress and performance—namely, parent-teacher conferences. Conferences that involve the evaluation of children's mathematics progress and performance can create opportunities and challenges for teachers, families, and students to positively shape mathematics learning and identity.

Enhancing communication practices that foster trust and mutual engagement among teachers, families, and students to support math learning is key in equity-focused teacher practice (Adams and Christenson 2000; Christenson 2004; Minke and Anderson 2003). We believe that parents and caregivers are underused resources as partners in mathematics learning and that building this partnership can greatly improve relationships between schools and the families that they serve.

Introducing Classroom Math Practices to Parents and Caregivers: Communicating Your Mathematics Vision

It is the beginning of the school year. You are getting your classroom ready, planning with your colleagues, reviewing your student lists, and preparing for the first day. You are outlining your goals, projects, and class management routines as you anticipate a new group of students. You are excited to meet your students and their families. Although all teachers develop welcome routines and practices designed to prepare students for the learning ahead, it is useful to take a closer look at specific classroom routines and practices that *introduce and welcome families to mathematics* in your classroom.

Think about how you communicate your vision for mathematics. What does mathematics learning look like in your classroom? What will be taught, and how you will assess and support students? As discussed in Chapter 3, your mathematics teaching identity—the totality of your beliefs and knowledge—shapes your mathematics vision and instructional practices. We now invite you to think carefully about how you share your vision for mathematics learning with your students' families when you employ two very common routine practices for communication with families: classroom newsletters and back-to-school nights. These are powerful tools for communicating your math vision to parents and caregivers, and you can leverage them to support math learning and teaching in your classroom.

Classroom newsletters—showcasing mathematics

For many teachers, a common means of communicating with families is the weekly classroom newsletter. The newsletter provides important information about classroom activities and upcoming events. Often these newsletters have special sections that showcase specific subject areas, such as reading, science, music, and mathematics. We invite you to rethink your classroom newsletter to emphasize mathematics teaching and learning in two important ways. We suggest two actions. First, within the first month of school, send a newsletter devoted to your mathematics learning goals for the year. Second, prepare a monthly or quarterly newsletter devoted to what is happening in mathematics in the classroom. Frequent communication that chronicles mathematics activities and student successes will help keep parents and caregivers informed and engaged in their students' mathematics learning.

To help with this process, we offer several design-related questions for mathematics newsletters:

- What is mathematics, and why is it important to learn?
- What are your goals for students as math learners?
- What does mathematics learning look like in your class?
- What does mathematics teaching look like in your class?
- How will you engage parents, caregivers, and the community in developing mathematics learning in your classroom?

- How will you help students with different mathematics preparation and levels of confidence succeed in mathematics?

- How will you help students with various ethnic, cultural, and socioeconomic backgrounds—some of whom are learning another language—succeed in mathematics?

Next, we discuss two examples of family newsletters that convey the teachers' mathematics vision for their classrooms. These samples are actual newsletters created by teachers whose real names appear in the discussion that follows. The first sample was sent to families of second graders early in the school year. The second sample was designed by a middle school mathematics teacher who sent out monthly math newsletters that focused on specific math topics. These newsletters offer examples of different ways to engage families as resources to support mathematics learning and positive identity development in the classroom. We consider these newsletters in turn in the discussion that follows.

Crazy for math

At the beginning of the school year, Mrs. Sabol, a new teacher working with second graders in an urban district in the Pacific Northwest, sent her students' families the two-page math newsletter shown in figure 10.1. Her school's student population is culturally diverse, with no identified racial or ethnic demographic group exceeding 30 percent. The school serves a sizeable population of multilingual learners (21 percent, compared with the state average of 8 percent), with the largest linguistic population speaking Spanish.

Mrs. Sabol's newsletter is informative and creative. She wants her students to be "crazy for math." Her newsletter positively conveys the equity-based mathematics teaching practices discussed in part 2 in many ways. It outlines very specific learning expectations that reflect a commitment to *going deep with mathematics, affirming mathematics learners' identities, leveraging multiple mathematical competencies*, and *challenging spaces of marginality*. On the newsletter's first page (Figure 10.1a), Mrs. Sabol conveys these expectations to parents and caregivers:

Your child and I will be fellow travelers as we test ideas, make conjectures, develop reasons, and offer explanations.

We also will learn that:

- Every student brings knowledge to our classroom.

- Ideas are the currency of our classroom. Your child's ideas have the potential to contribute to everyone's learning and will receive respect and response!

- Students must respect the need for everyone to understand their own methods and must recognize that there are often a variety of methods that will lead to a solution.

- Mistakes are an opportunity to learn. They give us a chance to examine errors in reasoning and contribute to everyone's learning.

Mrs. Sabol's Math News   **January 10, 2011**

Dear Parents,

I am excited to have your child in my math class! I plan to encourage a spirit of inquiry, trust and expectation in my classroom.

Did you learn math from lectures and worksheets? Were you taught to simply repeat back what your math teacher told you? Were you then expected to apply these concepts in your real life?

Math has turned around!

We will link home and school in our math classroom by building on experiences your student shares with others in our class. I will teach your student to do math like they would in the real world, starting with problems, dilemmas and questions they would encounter in life! Your child and I will be fellow travelers as we test ideas, make conjectures, develop reasons, and offer explanations.

We also will learn that:
- Every student brings knowledge to our classroom.
- Ideas are the currency of our classroom. Your child's ideas have the potential to contribute to everyone's learning, and will receive respect and response!
- Students must respect the need for everyone to understand their own methods and must recognize that there are often a variety of methods that will lead to a solution.
- Mistakes are an opportunity to learn. They give us a chance to examine errors in reasoning, and contribute to everyone's learning.

****My goal is to make your child crazy for math! Please complete the attached survey. With your help I can design math problems that connect with their life!*

Mathematically yours,

Mrs. Sabol

What is Math?

It is a science of concepts and processes that have a pattern of regularity and logical order.

Finding and exploring this regularity or order, and then making sense of it, is what doing mathematics is all about.

I Believe worthwhile math problems...

... allow for **connections** (to home and community)

... incorporate **multiple approaches and solutions** (using problem solving process and developing associated skills are often more important than the "right" answer).

... require **higher level thinking**

... facilitate **reasoning and communicating mathematically** (students need opportunities to talk about their math thinking).

Do you know?... "Researchers consistently find that the most important factor in school success is what they call "opportunity to learn."' If students are not given opportunities to learn challenging and high level work, then they do not achieve at high levels." -Jo Boaler (from What's Math Got to Do With It? Please see me if you would like the full article). Students in my classroom receive challenging, engaging, and high level work no matter what their confidence level.
My students work respectfully together and help each other!

(a)

Fig. 10.1. A math newsletter for families of Mrs. Sabol's second graders

Why is learning math so important?

We use math in every aspect of our lives in one manner or another. When we drive a car, we compute our speed and how many miles we have gone. We use math to balance checkbooks. Even musicians use math when playing instruments. You and your child use math when you:

- Go shopping
- Tip at restaurants
- Check the change you receive
- Measure ingredients
- Calculate time to do something
- Measure fabric to sew
- Read a clock
- Figure out batting averages
- Calculate mileage and gas money
- Buy lunch at school
- Divide pizza or other treats
- Measure weight
- Plan a lemonade stand
- Figure out the time somewhere else
- Calculate how many days until our birthday
- Play video games
- Measure distance
- Use a calendar
- Figure out the price of a sale item
- Decide when to set the alarm
- Use the telephone
- Count allowance
- Play board games
- Build something
- Calculate postage
- Figure out sales tax
- Watch football
- Budget money
- Read a thermometer
- Figure out cost of movie snacks
- Pay bills

Math Is Everywhere!

Help your child strengthen math abilities at home!

Everyday experiences at home can help your student learn math? Combining common sense with mathematical thinking helps your child learn! Look for times when math happens naturally.

For some ideas about finding math opportunities at home,
→ *visit: http://athomewithmath.terc.edu/* ←

(website available in Spanish and English)

MATH is FUN!

Learning through problem solving:

- Puts the focus on ideas and sense making
- Develops math confidence
- Teaches students that math makes sense
- Provides a context to build meaning for concepts
- Allows an entry point for a wide range of students
- Provides valuable data to assist me in helping students and keep parents informed
- Allows for extensions and elaborations
- Reduces discipline problems
- Develops mathematical power
- Is a lot of fun!

My goals for your student:

Gain confidence and belief in abilities.

Be willing to take risks and to persevere

Enjoy doing math

To help your student reach these goals, I will:

Build in success!

Praise efforts and risk taking!

Listen to all students!

(b)

Fig. 10.1. A math newsletter for families of Mrs. Sabol's second graders, *continued*

Mrs. Sabol is setting the stage for her students and their families to understand the mathematics learning and teaching expectations of her classroom. She conveys the idea that math learning is a journey that she is making *with* her students. Ideas, reasoning, solutions, and mistakes are all part of the mathematics learning in this classroom. Student contributions are valued and respected. The last two sentences in the "Do you know? ..." box at the foot of the first page (see Fig. 10.1a) reiterate this idea:

Students in my classroom receive challenging, engaging, and high-level work no matter what their confidence level.

My students work respectfully together and help each other!

These statements suggest a strong commitment to building a mathematics learning community in which all students will actively participate, contribute, and learn. The statements emphasize mathematical proficiency and agency to support students' mathematics learning.

Mrs. Sabol's newsletter also makes explicit how to tap into students' *multiple resources of knowledge* and everyday experiences while affirming their cultural and linguistic knowledge. "Math is everywhere," the newsletter states, providing examples of the various ways that families use mathematics in common activities such as shopping, cooking, sewing, budgeting, and playing sports and video games. It also identifies a bilingual website (English and Spanish) where parents can find more ideas on math activities that families can do at home.

A noteworthy component of Mrs. Sabol's newsletter is an enclosed bilingual questionnaire (Spanish and English) for parents and families to complete to help Mrs. Sabol design meaningful and relevant mathematics lessons for her students (see figure 10.2). She asks questions about family experiences, languages spoken, hobbies, traditions, work, and pets. The inviting tone of her statement about the questionnaire in the newsletter (*"With your help, I can design math problems that connect with their life!"*) and her use of two languages welcomes parents as partners and resources in supporting mathematics learning and teaching in the classroom. Mrs. Sabol clearly intends to engage parents in helping students learn mathematics with her. She obviously wants to deepen her students' enjoyment of mathematics and assist them in seeing its relevance and connection to their world.

The power of proportion

In middle school, mathematics learning can strengthen or diminish students' confidence in their ability to solve problems. "When am I ever going to use this?" is one of the most frequent questions raised by middle school students about mathematics. To preemptively address this common question, one middle school mathematics teacher, Ms. Le Sage, created a monthly mathematics newsletter for her students and their families, tied to specific math topics such as scale, ratio, and proportion.

An important feature of Ms. Le Sage's newsletter is a monthly calendar with scale-related mathematics problems and activities that families can investigate (see the sample in figure 10.3). This monthly calendar feature complements the specific definitions, examples, and ways to solve problems involving ratio and proportion studied in the classroom. The newsletter gives parents and caregivers an opportunity to learn in a variety of contexts.

In the body of the newsletter (see figure 10.4), Ms. Le Sage suggests various ways for families to engage in mathematics discussions that tap into their experiences and

Make math come alive for your child!

Knowing about your child's background and knowledge will help me to design math problems that relate to their lives! Please complete this survey and return it to me as soon as possible. Help to make your student crazy about math!

1. Do you have a family pet or pets? If so, please list your pets and their names. _____
    _____
    If you do have pets, would you be willing to bring your pet for a visit to our classroom? _____
    _____
    _____

2. Does your child have experiences with grandparents or aunts and uncles who have interesting stories or experiences they could share with our class? _____
    _____
    _____

3. In what ways are you and your child proud of your home? Please use the space below to tell about your family's rituals and traditions including food, music, birthdays, holidays, and the work that family members do: _____
    _____
    _____
    _____
    _____
    _____
    _____
    _____

4. Do your family members share any hobbies?_____
    _____
    _____

5. What languages are spoken at home?_____

6. How many siblings does your child have? Brothers?_____ Sisters?_____

7. Volunteers are welcome and needed in our classroom! Do you or any other family member have time to contribute, or any interesting hobbies or activities you would like to share with our class? _____
    _____
    _____ If you
    would like to help out or share in our classroom, what is the best way to contact you?_____
    _____

Thank you for completing our survey!

Watch for news about how your child's experiences at home contribute to math problems in class!!

(a)

Fig. 10.2. A bilingual questionnaire in (a) English and (b) Spanish enclosed in Mrs. Sabol's newsletter to families

¡Haga que las matemáticas sean parte de la vida de su hijo/a!

Saber sobre las experiencias de su hijo/a me ayudará a diseñar lecciones de matemáticas que sean más pertinentes a su vida. Favor de contestar las preguntas abajo, y devuélva la encuesta lo más pronto que sea possible. Asi me ayudará a hacer que las matemáticas sean algo emocionante para su hijo/a.

1. ¿Ustedes tienen una mascota en la casa? Si es así, haga una lista de sus mascotas y sus nombres.
    _____.
    ¿Si ustedes tienen mascotas en la casa, estarían dispuestos a traerlos al salón de clases para una visita?_____
    _____

2. ¿Si hijo/a tiene abuelos o tíos/tías que podrían venir a nuestro salón a platicar con los estudiantes sobre sus experiencias? _____
    _____
    _____

3. Me gustaría aprender más de su familia. Utilice el espacio abajo para describir algunas de sus tradiciones importantes. Por ejemplo, cómo celebran los cumpleaños y días de fiesta, y cuál es la comida y la música que más les gusta. También describan los trabajos que hacen los miembros de la familia. _____
    _____
    _____
    _____
    _____

4. ¿Sus familiares tienen pasatiempos? Descríbalos aquí. _____
    _____
    _____

5. ¿Cuáles idiomas hablan en su casa? _____

6. ¿Cuántos hermanos/as tiene su hijo/a? ¿Hermanos? _____¿Hermanas?_____

7. Las personas voluntarias son totalmente bienvenidas en nuestro salón de clase. ¿Usted o algún miembro de su familia tiene tiempo para ayudar en la clase, o para platicar con los estudiantes de sus pasatiempos? _____
    _____¿Si a
    Ud. le gustaría ayudar en nuestro salón de clase, cuál es la mejor manera de contactarle?_____

Gracias por contester estas preguntas.

Esté atento a como las experiencias de su hijo/a contribuyen a las lecciones de matemáticas.

(b)

Fig. 10.2. A bilingual questionnaire in (a) English and (b) Spanish enclosed in Mrs. Sabol's newsletter to families, *continued*

household funds of knowledge, encouraging them to use these as mathematics resources for deepening students' mathematical understanding (Civil 2007; González et al. 2001). Family discussion prompts appear in the "Mealtime Moment" section of the newsletter, and specific questions in the "Project in Progress" section enable parents and caregivers to ask their children about the class project currently underway—the restaurant project, which is related to scale and ratio and engages the students in floor plan design, recipe modifications, and seating arrangements.

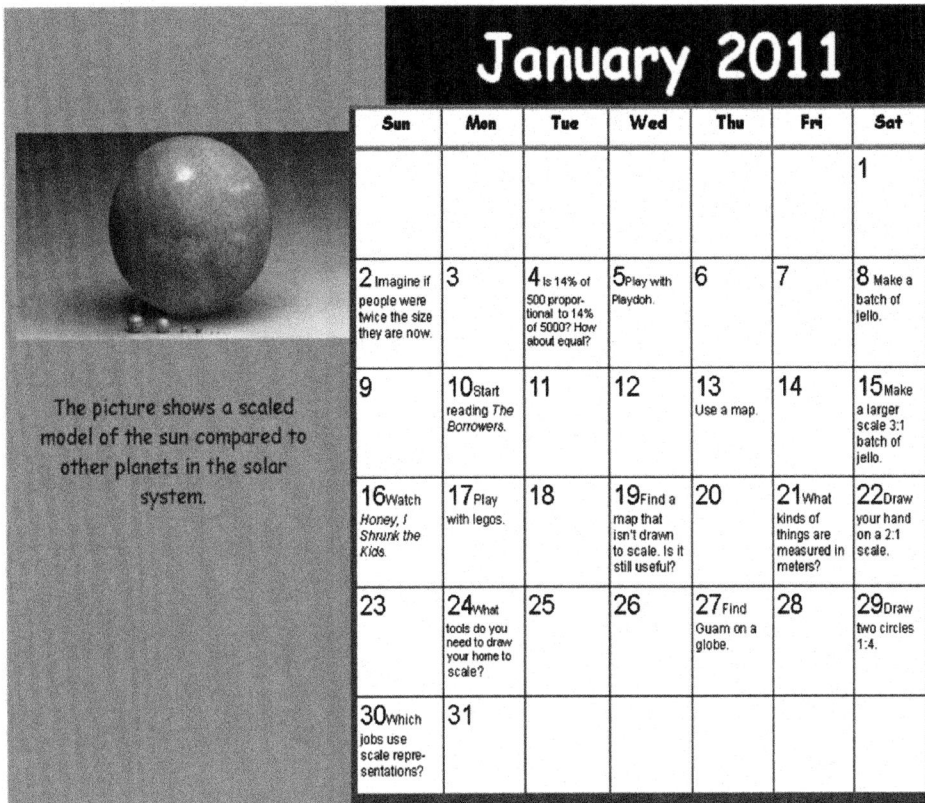

Fig. 10.3. A sample calendar from a middle school teacher's monthly math newsletter

It is important to note that Ms. Le Sage teaches middle school mathematics at a tribal school. Tribal schools are independent schools run by Native American tribes for Native American students, to support cultural and linguistic sustainability of Native American heritage (some schools are affiliated with the Bureau of Indian Affairs and receive federal aid). One of Ms. Le Sage's goals is to connect her students and their families to Native American scientists. The newsletter features a biography of Wilfred Foster Denetclaw, Navajo and university biologist. Dr. Denetclaw offers advice on being educationally successful and discusses his research interests. Ms. Le Sage selected this biography to show a cultural role model in mathematics and science to affirm her Native American students' mathematical identities. Although this may seem like a small connection, providing families with such examples produces additional sources of support to affirm students' multiple identities and continue their mathematics education. (Biographies of mathematicians and scientists of color are available from organizations such as the Society for the Advancement of Hispanics/Chicanos and Native Americans in Science [www.sacnas.org] and the Benjamin Banneker Association [www.bannekermath.org].)

Both sample newsletters offer important resources to extend mathematics learning into family routines, enhancing opportunities to discuss mathematics with students at home. These are not typical "monitor homework" suggestions but authentic, creative ways to connect classroom mathematics learning with the students' home lives.

January 2011

Miss Le Sage • 253-435-3000 Ext. 3142 • Desiree@leschischools.org

Math Beat

Big Name of the Math Dr. Wilfred Foster Denetclaw - *Zoologist*

"My life has been shaped by the fairly traditional Navajo lifestyle I was raised in. It taught me to be respectful, and to be responsible...I did not do well my first time in college. I had been at the top of my class in high school, but I was not prepared for coursework at a major university. I had not learned how to write well and I did not have an understanding of mathematics or science that other students possessed. I did not know how to apply my time well for studying [or] how to use the educational resources at the university. I overcame these problems, first, by keeping my desire for a college education alive, second, by having my parents support my interest in college, and third, by getting serious about learning. Since, I have been involved in many interesting research projects... For example, I have been looking at the earliest stages of skeletal muscle development using the chicken embryo as a model for understanding muscle formation in humans."

Focus of the Week SCALE

Understanding scale is important because it engages proportional reasoning skills. Have you ever tried making a small batch of a recipe before cooking that recipe for a large group? You tested a small scale of the recipe! **Scales change the size of an entire representation by the same amount.** They can be used to plan, practice, or visualize for a variety of reasons. In NASA, scale models were often used because building a large scale spaceship incorrectly would be very costly. Scales save money! Helpful concepts in math scaling are multiplication, fractions, unit conversions (ex: feet to inches), and ratios.

Mealtime Moment

Why is it important for All the items on a map to be drawn to the same scale? What if you drew the roads on a different scale than city boundaries?

Scale is the ratio between the size of something and the representation of it.

Project in Progress

Currently, we are creating a scaled-down floor plan for our class restaurant that we have named _____. Ask your children questions about their design and bring up restaurant floor plans if you go out to eat! You can even talk about how your home dining area is set up. Then, we are going to discuss scaled-up recipes for making many servings of a single recipe. You can practice this one at home.

Algorithm Example

A model airplane is built on the scale of 1:18. If the wingspan of the plane is 30 feet, the wingspan of the model is how many inches?

Since the wingspan of the airplane was given to us in feet we must change it to inches. 30 feet = 360 inches.

Setting up the proportion:

$$\frac{model}{actual} = \frac{1}{18} = \frac{x}{360}$$

$$\frac{18x}{18} = \frac{360}{18}$$

$x = 20$ inches

Student Spotlight

_____ noticed that last year's yearbook used scale with some pictures, and didn't use scale with other pictures. She pointed out that the pictures that weren't scaled made people look different than what they really are. Great observation _____!

What do YOU want your child to learn in math? I'd love for you to talk about your ideas for projects, lessons, field trips, guest speakers, etc. Never forget, you are an icon for your child.

Fig. 10.4. A math newsletter for families of middle school students

Curriculum Back-to-School Nights

Bringing parents and teachers together, the annual back-to-school night has become a hallmark of the school year. This common event offers teachers their primary opportunity to showcase to parents and caregivers what their children will be learning during the school year. It is often the forum where teachers describe their learning expectations and curriculum foci. We encourage you to think about how you introduce your vision for mathematics to your parents and caregivers at this event.

In the following vignette, the parent of a third grader at a public elementary school recounts her experience at back-to-school night. Her account suggests how a teacher's enthusiasm—or lack of enthusiasm—for a subject gets communicated. Particularly noteworthy is how the teacher's presentation positions mathematics in relation to other school subjects:

We all sat at our child's desk, each family member (moms, dads, grandparents, sisters, aunts) eagerly wanting to know what was going to be learned. The teacher had a PowerPoint presentation projected on what she called the "smart board." The teacher enthusiastically described the new reading program ("in third grade, we *read to learn* rather than *learn to read*"), the "fantastic" science projects (e.g., wetlands, solar system, salmon migration), and "really interesting" social studies projects (e.g., immigration and Native American history). However, when the slides turned to mathematics, the teacher's enthusiasm visibly waned as she gave the name of the curriculum used and gave the standardized testing dates and plans for test preparation. She spent less than a minute talking about math, with no examples and no smile. I wondered, does she even like teaching math?

The teacher's presentation exposes the limitations of her vision for mathematics. In contrast with her treatment of the other subjects, she discussed mathematics solely through references to the textbook and standardized tests, with no effort to highlight a set of mathematical competencies and relationships integral to students' problem solving or expanding ability to make sense of the world (Gutstein 2006; Schoenfeld 1992). The presentation did not associate mathematics with adjectives like *fantastic, interesting,* or *fun.*

A key consideration for teachers about back-to-school night is the *image of mathematics* that is projected to families. Back-to-school night is an opportunity for you to clearly communicate the mathematical ideas and processes that students will be studying and why those topics are important. It is an opportunity to invite parents and caregivers to participate in your mathematics vision for their children in new and exciting ways. Refining your message about your vision of mathematics as informative, creative, and positive can help set the stage for a positive year of mathematics learning. The following are some key elements to consider including in your back-to-school presentation about mathematics:

- Enthusiasm for the subject—communicating why math is important for students to learn (beyond being measured by standardized tests)
- Ways that you will nurture a positive math identity
- Ways that you will help strengthen students' mathematical proficiency and agency
- Your favorite mathematics topics to teach for this grade level and why you like these best
- The strengths of the mathematics curriculum and ways that you will enrich the students' mathematics experiences
- Special math projects (perhaps tied to science units or community service projects) that your students will do
- Ways that you will communicate your students' math progress to families

- Ways that parents and caregivers can partner with you to support math learning (you might suggest family games—such as dominoes, chess or checkers, Parcheesi, cribbage, Connect Four, or spades—and you could actively point out ways that families can use mathematics in everyday events, such as estimating grocery bills, cooking, measuring fabric, completing repair projects)

To enrich this experience, you might consider distributing a mathematics survey to family members to obtain information that will help you build on students' multiple knowledge bases and experiences. Like Mrs. Sabol's newsletter survey, your survey might give you information that you could use as another resource for your mathematics lesson planning. You might also distribute a mathematics resource list with activities and websites for families to visit after school or on weekends. Parents and caregivers want to know how they can help their child to do well and enjoy learning mathematics. The key is to share your vision and how you want to work with families to engage their children in mathematics learning that promotes a positive mathematics identity and gives them a strong mathematical knowledge base.

How Is My Child Doing in Math? Communicating Progress and Performance

Parents and caregivers learn about their children's math performance in several other common ways, such as during parent-teacher conferences and through artifacts of performance—namely, report cards and unit and chapter assessments. These are *vulnerable contexts* for parents and caregivers as they learn from another adult—an instructional expert—how their child is doing academically, socially, and emotionally. These interactions have the power to influence how parents and caregivers will perceive a child's identity and success or failure in mathematics. Furthermore, these interactions introduce tensions and can create opportunities to increase or decrease trust between parents, caregivers, and schools (Adams and Christenson 2000; Christenson 2004; Minke and Anderson 2003). Thus, the term *vulnerable* is fully warranted.

Consider the following scenario involving a Latina mother who recalled her recent parent-teacher conference about her third-grade daughter, Xiomara:

The teacher began listing all her "weaknesses," from struggling with memorizing her addition, subtraction, and multiplication facts to her inability to solve word problems. The teacher said that if Xiomara did not know her facts she would be unable to solve more complex problems in fourth grade. The teacher showed examples of mistakes our daughter made on the most recent unit review test. Confused, I asked the teacher why she gave Xiomara a score of "3–meeting grade-level standard" in mathematics on her most recent report card. The teacher responded by saying, "She seems to pull it together for the tests." I asked, "Is there anything Xiomara can do in math?" She simply said, "No. Xiomara really struggles." Then the teacher described all the "interventions" she gave our daughter during math lessons, such as extra tutoring from herself or a parent volunteer and extra time to complete assignments. We were unaware of this extra help that was being given to our daughter, and we asked why we had not been told earlier of her struggles since this was

December. The teacher replied, "Don't you see the graded assignments and tests?" We walked away from this conference very upset. Our daughter had no mathematical strengths, only weaknesses. What does that mean? How can we help? There were no resources given to us.

According to the teacher, this child was not making adequate mathematical progress, even with several interventions. The teacher's intent was to bring this concern to the parents' attention during this critical school event—the parent-teacher conference. Unfortunately, the conflicting pieces of assessment data, the focus on the child's weaknesses, and the lack of direction to parents to help their child improve rendered this interaction very tense and unsupportive. The parents had now acquired a negative image of their daughter as a mathematics learner without being given any resources to address her needs.

Because parent-teacher conferences are critical venues for establishing and maintaining the trust required for effective school-family partnerships, it is essential to reflect on these interactions and their impact on parents' and caregivers' views of their children as mathematical learners. The images, evidence of progress, and mathematical strengths and needs that are communicated to parents can affect how they see their children as learners of mathematics and can shape at-home interactions with math.

To help facilitate productive and positive discussions about a child's mathematical progress and developing mathematics identity, we propose a basic template for parent-teacher conferences. The template, which appears in figure 10.5 (and is also available at nctm.org/more4u), communicates the core mathematics learning goals of a teacher's mathematics vision and reflects a holistic and balanced evaluation of a child's mathematics progress with a mathematics action plan (MAP) for learning support. The structure of parent-teacher conference template reminds everyone that learning mathematics is important and that everyone can do it with encouragement and instructional assistance.

The template makes the child's mathematical strengths and needs explicit. Regardless of the grade on the report card, every child has mathematical strengths and growth areas. A child's mathematical strengths must be affirmed for the child to maintain a positive mathematics identity. The student's mathematical growth areas must also be addressed for the student to move forward, developing mathematical knowledge and practices. Further, the MAP provides a space to record specific ways in which the student, parent or caregiver, and teacher can work together to support and enrich the child's mathematics education.

We envision teachers using this conference template in conjunction with student work. Furthermore, the MAP should include a way that the student, teacher, and parent or caregiver can address a specific need or build on a specific strength. For example, in the case of Xiomara, who is struggling with her addition, subtraction, and multiplication facts, the MAP might put the following action items into place:

Student Action:

- Be positive. Remember, you can do this!
- Solve an online math puzzle or play an online math game. Check out NCTM mobile apps with games like "Product Game" (https://www.nctm.org/Classroom-Resources/Illuminations/Interactives/Product-Game/) to practice math facts.
- Spend twenty minutes three times a week practicing/playing the online game.

Parent-Teacher Conference

Math Vision: Core mathematics learning goals affirming why mathematics is important for children to learn (e.g., working together to learn math together!). A statement that every child can be successful in mathematics with the proper instruction, support, and encouragement.

Student: Teacher:

Mathematics Progress

Mathematics Strengths (Areas of Strength):

Mathematics Needs (Areas for Growth):

Mathematics Action Plan (MAP)

Student Action:

Teacher Action:

Parent Action:

Fig. 10.5. A feedback template for parent-teacher conferences

Teacher Action:

- Have Xiomara set up a hundreds chart to color-code facts that she knows. Talk with her about how to use these facts, along with strategies like doubling, to help her with facts that she is still working on.

- Set up a station lesson or menu lesson for students, with math facts as one of the required items. Other items might include math array games and patterns.

- Introduce Number Talks (Humphreys and Parker 2015; Parrish 2010), structured ten- to fifteen-minute whole-class mental math discussions of number and operations, to reinforce skill with mental math and in decomposing numbers.

- Set up study groups for students to learn how to work and study together. Pair Xiomara with a student who has some efficient math strategies for multiplication facts.

Parent Action:

- Play online math games with Xiomara to practice math facts.
- Play family games like cribbage or dominoes to practice addition and multiplication.
- Work with Xiomara to create a "favorite number" book with photographs, stories, and drawings representing Xiomara's favorite number in many ways. Practice math facts with Xiomara by making calculations that use different numbers and operations to arrive at that number (Coates and Thompson 2003)
- Take a math walk or play "Math I Spy" at home, pointing out instances of multiplication arrays in the home, such as egg cartons, cupcake tins, windowpanes, candy boxes, shelves, and so on. Have Xiomara tell you the multiplication fact represented by the array.

Although the teacher should complete this template before the conference, part of it could be coconstructed by all participants during the meeting. For instance, the MAP can serve as a brainstorming tool and a memorandum of understanding for the student, teacher, and parent and caregiver. Moreover, for older students, contributing ideas about how they will strengthen their own mathematics learning beyond the classroom door can be a way of supporting a sense of individual agency. The ultimate goal is to have everyone walk away from such a meeting with an understanding of the child's mathematical strengths and growth areas and a specific plan that shows how mathematics learning will be supported at school and home. We also encourage the teacher to follow up with parents and caregivers and the student about progress on the action items. Communication with parents or caregivers might be in the form of a quick e-mail or phone check-in to let them know how you have implemented your action items with their child. This contact can open additional dialogues about progress or more support resources if needed.

Initially, creating a summary of strengths and growth areas and a MAP for each student will be time-consuming. However, we believe that the investment of this time will provide teachers with a deeper understanding of how to support each child more effectively; give parents, caregivers, and teachers alike concrete ways to affirm to the child that they can learn mathematics; and promote a culture of mutual trust and respect that can strengthen family partnerships that are needed to help children learn mathematics.

Conclusion

This chapter has examined some routine practices and contexts that allow teachers to communicate with parents and caregivers about the mathematical learning environment in which their child participates. Although routine, these practices can be powerful in conveying mathematical ideas, learning goals, instructional strategies, and student progress.

Classroom newsletters and back-to-school nights can be tailored to promote a positive image of mathematics learning—an image that is exciting to both students and their families. This is important for two reasons: to combat the negative images that many adults have of mathematics and to provide clear direction and resources for families to connect with mathematics in new and inviting ways. In addition, these formats can also provide teachers with important information through surveys and conversations, and this information can be integrated into mathematics lessons to maximize interest, relevance, and participation.

Parent-teacher conferences are another regular practice with tremendous implications for how parents and caregivers view their children as mathematical learners. The conference template that we have presented offers teachers a way to reaffirm their vision for mathematics learning in the classroom, engage in a more substantive discussion of a child's areas of mathematical strength and need, and give clear guidance through the mathematics action plan so that everyone—the child, the teacher, and the family member or caregiver—can take responsibility for the child's mathematical learning. These kinds of actions can go a long way toward building families' trust in the teacher as a true partner, committed to helping their children learn mathematics with confidence.

DISCUSSION QUESTIONS

1. How do you communicate your math vision to families?

2. How do you communicate a child's mathematical strengths and growth areas to parents and caregivers?

3. In light of the examples presented here, how might you strengthen your communication practices about mathematics with your families?

4. Think about a student who struggles or needs enrichment. What would a mathematics action plan (MAP) look like for that student?

Chapter 11

Partnering with Families and Communities to Support Children's Equitable Mathematics Learning

Parents—including all adult caregivers who play parental roles in children's lives, such as grandparents, guardians, foster parents, aunts, uncles, and so on—often underestimate the power of their words and actions on the children in their care:

> I didn't realize how much I was influencing my child's math performance by my comments that I was no good at math as a kid. I had to do something to change both of our attitudes. (EQUALS/FAMILY MATH parent)

It is well established that children whose parents are involved in their children's education experience higher academic achievement regardless of family income, level of education, or cultural background (Epstein 1996). This has also been shown to be true regarding their mathematics education (Desimone 1999; Epstein 1984; Yan and Lin 2005). However, engaging families—including parents and caregivers—in ways that positively influence their children's mathematics learning can be challenging, given the content standards and accountability demands currently imposed on teachers and students. For example, federal requirements regarding student and school benchmarks for growth, along with the threat of state-imposed sanctions, have heightened scrutiny and control of classroom instruction, content, and pedagogy. Consequently, meeting district and state benchmarks consumes the majority of teachers' instructional planning time.

Furthermore, many parents and caregivers have not experienced success in mathematics, a situation that can lead to math avoidance or anxiety that they unwittingly pass on to their children. These mathematics experiences can also prevent parents and caregivers from being involved in or advocating for their children's mathematics education. Parents who have experienced success in mathematics or who can afford to provide additional resources to enhance their child's mathematics learning are able to play a more active and influential role in their children's mathematics education (Martin 2000, 2006; Remillard and Jackson 2006). Consequently, their children may often feel more confident and competent in mathematics.

We believe that if teachers are to engage students who have been marginalized in mathematics education and increase their success, they must engage *all* parents in meaningful discussions about mathematics. Studies have shown that mathematics reform promotes a perspective on mathematics that is often unfamiliar and inaccessible to parents who have not had opportunities to learn mathematics in a conceptual way (see, for example, Remillard and Jackson [2006]). Furthermore, African American, Latinx, and low-income parents often experience more challenges in relation to their children's education because they are frequently perceived as uncaring or disengaged (DeCastro-Ambrosetti and Cho 2005). They are often seen as obstacles rather than resources for their children and are subsequently left out of critical conversations about mathematics reform efforts and classroom instruction (Martin 2000, 2006; Peressini 1998; Remillard and Jackson 2006).

Creating and sustaining meaningful ways to interact with parents and caregivers about their children's mathematical experiences in and out of school are paramount in helping children develop positive mathematics identities and achieve success in school. As discussed in Chapter 10, school systems have established routines and structures, such as back-to-school nights, classroom newsletters, and parent-teacher conferences, to communicate with and, to a certain degree, partner with parents and caregivers to enhance students' mathematics success. School systems and teachers, however, may need to forge new pathways and enlist parent and caregiver assistance in contexts beyond the classroom setting to achieve success with all students. Creating new pathways takes additional effort and time—both scarce commodities, given the demands of teaching and the complex lives of families. Making any change in one's teaching practice takes commitment and perseverance, but once new practices are internalized and the benefits become apparent, the practices become routine.

This chapter provides examples of family-school partnerships focused on mathematics. The vignettes that follow demonstrate how partnering with parents, caregivers, and community groups from out-of-school settings can provide children with additional opportunities to deepen their conceptual understanding of mathematics and develop positive mathematics identities. We showcase the FAMILY MATH program (EQUALS, University of California, http://www.lawrencehallofscience.org/equals/aboutfm.html) and the Math and Parent Partners project (MAPPS; University of Arizona, https://mappsua.wordpress.com) as models for parent and caregiver engagement that transcend traditional school–home boundaries to create mathematically rich environments that can support mathematics instruction that occurs in classrooms.

We also discuss mathematics partnerships formed with community centers and faith-based organizations to provide additional ways to engage children in mathematics and affirm their mathematics identities. We invite teachers to reflect on their assumptions about parents, caregivers, families, and communities and the roles that each plays in their students' mathematical development and then consider how to connect those previous assumptions with their new insights to enrich current partnerships and develop new ones to engage families more fully as resources for mathematics teaching and learning.

Using Family-Focused Mathematics Education Programs

Like Janelle's mother in the following dialogue, most parents and caregivers want to help their children succeed in mathematics:

Janelle:	Mom, I'm done with my homework. Can I play a game on the computer?
Mom:	Just a minute. Let me see your homework. You didn't finish, Janelle. What about the fraction problems?
Janelle:	[*Sighs*] I got stuck on the last set of problems. Mrs. Schneider told us the rule. I know I need to make the bottom numbers the same. I just can't remember how.
Mom:	It's OK. Let me show you another way to compare fractions. Remember the number-line activity we did during the FAMILY MATH night at school? You were good at showing me $1/2$ and $1/4$ on the number line. Let's use that method to compare fractions in this problem.

Because Janelle's mother participated in a math program designed to assist parents and caregivers in helping their children at home, she now has the confidence, knowledge, and skills to support Janelle's mathematics learning and can truly partner with her child's classroom teacher. Many programs are designed to provide parents and caregivers—particularly those whose children have not achieved mathematical success—with the tools to help their child be successful in mathematics.

The FAMILY MATH program

FAMILY MATH, an EQUALS program, one of the earliest family-focused mathematics education programs. EQUALS is a mathematics and equity program based at the Lawrence Hall of Science in Berkeley, California. FAMILY MATH focuses on creating access to mathematics for all students, particularly those marginalized in mathematics education. The program focuses on providing *all* parents and caregivers with the tools and resources to support their children's mathematics education. At a site meeting, Virginia Thompson, FAMILY MATH founding director, offered a description of the program:

> Teachers who participated in our EQUALS workshops on mathematics and equity wanted activities, other than worksheets, that they could give parents to help them support their children's mathematics at home. Parents knew how to help their children increase their reading comprehension skills, even if they weren't avid readers themselves. Reading at home was a comfortable, supportive interaction both parents and children enjoyed. The context was not the same for mathematics.

The FAMILY MATH program helps parents and children learn mathematics together. A major goal of activities in the program is to create that same sense of routine enjoyment in mathematics activities that many families experience in reading activities and to help empower students mathematically by providing tools and resources that parents can use at home. Additional program goals include the following:

- Raise parent and caregiver awareness of the role that mathematics plays in their children's future

- Provide parents and caregivers with strategies and materials to support their children's math learning at home

- Provide all parents and caregivers with the understanding and tools to advocate for their children's math education

- Increase parents' and caregivers' conceptual understanding of mathematics content

FAMILY MATH format

FAMILY MATH activities focus on understanding key mathematics concepts in the context of a game or an investigative format. These formats foster positive family relationships with school mathematics. Parents and caregivers can deepen their children's mathematics learning in a less formal context.

Consider, for example, a FAMILY MATH activity at the middle school level, "Let's Go to the Movies!" (Mayfield-Ingram and Ramirez 2005). This algebra activity is designed to help children practice combining like terms, a common algebraic procedure. The context of the activity is a school project: Century Hilltop Movie Theaters has invited students at Martin Luther King Middle School to preview twelve new kid-friendly movies. The teachers want the students to write movie reviews, and they decide to sort the students into review groups by having them identify algebraic "like terms" ($6a^2b$, a^2b; $7xy$, xy; y^2, $-5y^2$). Each student receives a "movie ticket" with an algebraic term. At the FAMILY MATH class, each family team sorts the movie tickets into like-term groups. The FAMILY MATH class leader models the process for sorting the term groups by holding up two tickets and asking the teams to compare them. Are they for the same movie? How do the team members know? How are the tickets alike or different? The family teams then work together to sort the tickets. Parents and caregivers mimic the questioning strategies of the class leader, a technique that helps everyone develop inquiry habits of thinking for solving mathematics problems. These habits help parents and caregivers assisting with homework at home as well as children working on classroom assignments at school.

This work with algebraic terms on movie tickets is preceded by an activity called "Do I Belong?" In this activity, families shuffle cards showing algebraic terms such as $18b$, $18c$, x^2, $5x^2$, and $-12x^2$. They discuss whether the cards show the same variable and whether the exponent is raised to the same power. The FAMILY MATH leader facilitates a whole-group discussion about exponents. What does the number represent? How are $2x$ and x^2 different? The leader selects a family team to record the examples (such as $x + x = 2x$; $x \cdot x = x^2$) the group gave on posterboard on a wall in the room (Mayfield-Ingram and Ramirez 2005).

These activities are designed to promote "math talk" among families to deepen student learning, support parent and caregiver learning, and engage families in mathematical ideas. Participating in these types of family-focused activities in a relaxed atmosphere provides more opportunities for a student to grapple conceptually with an important algebraic topic. It also enables parents and caregivers to learn or reacquaint themselves with concepts so that they can help support their child's mathematics education.

In the FAMILY MATH class environment, it is not only acceptable but also required to ask questions, work together, and model different strategies. The classes are held in a variety of settings, including libraries, churches, and parent-teacher association meetings, and in a variety of time periods, such as Saturday mornings, early evenings, and summer and spring breaks. Moreover, most materials are translated into Spanish, Mandarin, and Swedish. Parents and caregivers often team with teachers to facilitate classes that create an experiential bridge that they can cross and recross many times throughout the year to share information to support a child's mathematics understanding.

FAMILY MATH outcomes

By participating in FAMILY MATH classes, families can obtain the tools to address the tensions that they often see at home in their child's response to mathematics assignments. "I now see how my child thinks mathematically" is a typical comment from participating parents, who are then able to take this knowledge and apply it when working with their child at home, as in the following example, which involves fraction comparison and continues the dialogue at the beginning of the chapter between Janelle and her mother:

Mom:	OK. Janelle, we know they are fractions. Are they greater or less than 1?
Janelle:	Less than.
Mom:	How do you know?
Janelle:	Because the numerator is less than the denominator.
Mom:	Great! So how much of the number line do we need to draw?
Janelle:	[*Thinks a bit before responding*] Just the part from 0 to 1.
Mom:	Right. OK, let's draw some number lines.

Janelle's mother proceeded to draw number lines for each of the fractions. She started to draw the intervals but remembered the FAMILY MATH class discussion about the importance of letting children do the thinking for themselves. The parent and caregiver role is not to tell children *how to* but to help them *think through* what to do. So, instead, Janelle's mother handed her daughter the pencil to let her draw and label the fraction intervals. She asked, "How can you tell which one is the larger fraction?" Janelle responded, "The one that covers the most distance on the number line?" "Yes," her mother replied. "Why don't you finish the rest of the problems on your paper. Tomorrow, you can ask Ms. Schneider to show you the rule again for finding common denominators. It's helpful to know the rule also because it's sometimes quicker and more efficient than drawing a number line." Janelle said, "OK, but I'm glad you knew another way to solve the problems." Her mom responded, "Me, too. We can practice a few more tomorrow night, using the rule." (See figure 11.1.)

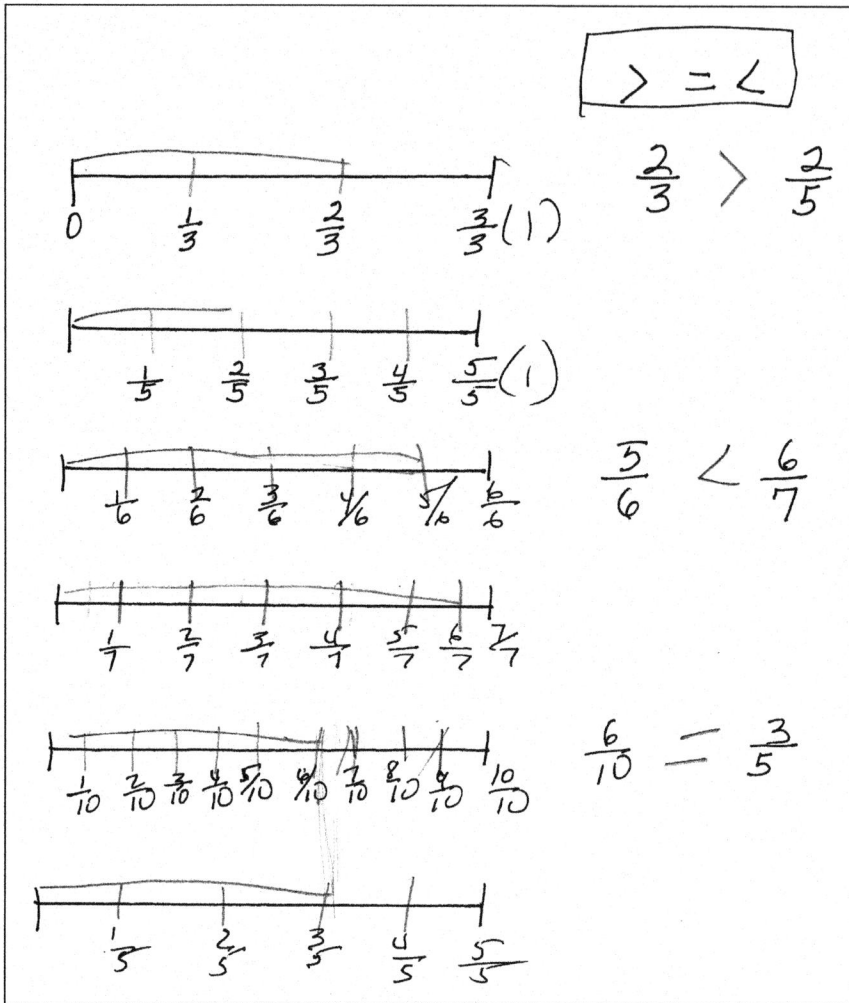

Fig. 11.1. Janelle's number line activity

These examples demonstrate how parent-child interactions can reinforce the equity-based practices that strengthen mathematics learning and support students' positive mathematics identities. For example, Janelle's mother was able to reinforce Janelle's conceptual understanding of a fraction as a number on a number line. Janelle could then use the structure to visualize and compare fractions. Teachers in earlier grades often use visual models to introduce fractions, and this was probably how Janelle encountered them earlier. In subsequent grades, however, many teachers rely on textbooks that quickly move to the standard algorithm for comparing fractions. Janelle's mother is tapping into previous mathematical knowledge of visual models of fractions to help Janelle make sense of these new fraction problems.

By using multiple representations—in this case, reconnecting Janelle to the visual model as another method to compare fractions—and encouraging Janelle to discuss the procedural rule with her mathematics teacher the following day, Janelle's mother helps to deepen her conceptual understanding of fractions. In this way, Janelle's mother is affirming her daughter's identity as a mathematics learner by helping Janelle practice perseverance with mathematics through coaching instead of explicitly doing the math problem for Janelle.

Allowing parents to experience standards-based mathematics tasks and activities can be much more beneficial than routine methods of disseminating mathematics reform to parents and caregivers. This will become even more important with states' implementation of content and process standards such as of the Common Core State Standards for Mathematics (National Governors Association Center for Best Practices and Council of Chief State School Officers 2010). CCSS-M is designed to increase student understanding and addresses both the content progressions necessary for student development and greater conceptual understanding. In addition to the Standards for Mathematical Content, CCSS-M includes the Standards for Mathematical Practice, which identify the learning habits that students should exhibit as they engage in mathematical problem solving. These practices include learning to persevere, modeling with mathematics, and constructing viable arguments. Such practices mirror other state's process standards such as the Texas Essential Skills and Knowledge standards (Texas Education Agency 2012).

When parents and caregivers understand how and why teachers engage in instructional practices such as questioning to elicit children's thinking or using models or drawings to represent mathematical concepts, the family dialogue at home is transformed. The focus on "getting the right answer" is replaced by a focus on "helping my child understand the solution." Teachers have limited instructional time and many concepts to address. If we can provide additional avenues for students to engage in rich mathematics learning outside the classroom, we can create new opportunities for teachers to go deeper in developing concepts inside the classroom. Programs like FAMILY MATH can become important components in a teacher's mathematics program design.

Math and Parent Partners project

Another program that creates partnerships between families and teachers for student mathematics success is the MAPPS project. Based on the funds of knowledge work in mathematics by Marta Civil and her colleagues, MAPPS explicitly rejects deficit views of families as lacking intellectual resources to help students learn mathematics. Instead, MAPPS views families as "intellectual assets" and seeks to support families in leveraging those assets by exploring concepts and skills that underlie the mathematics that children learn in kindergarten through grade 8 and connecting those concepts and skills with home and community activities (Civil and Andrade 2003; Civil and Bernier 2006). The program offers five minicourses (Thinking about Numbers; Thinking about Fractions, Decimals, and Percents; Geometry for Parents; Thinking in Patterns; and Data for Parents), spread out over a single semester and focusing on a mathematics theme.

Parents and teachers take courses together, strengthening the family-teacher relationship. The topics are presented through hands-on materials. Parents and teachers work in small groups and present their solutions to the whole group. Program evaluation has shown that teachers learn about parent mathematical understandings, problem-solving strategies, and ways of explaining topics to their child. Teachers also gain mathematics knowledge for teaching and discover how to secure parent involvement in a way that enhances student learning. For their part, participating parents view their relationships with teachers in new ways, seeing themselves as equal partners in the mathematical development of their children. A Latina mother who cofacilitated work-shops with teachers described this transformation:

> It was hard in the beginning to work with the teachers. "They are the best." They don't give you the opportunity that you may know more or bring other ideas. Now we are more equal. Before [with her hands she indicates parents in the team were at a lower level than teachers], but now [she indicates they are at the same level]. Now they rely on me, they check with me, they make you feel that you are important to them. (Civil and Bernier 2006, p. 327)

The MAPPS program seeks to help parents and teachers to work together to un-derstand mathematics concepts and learn pedagogical strategies for communicating the content to children. After taking the minicourse Thinking about Fractions, Decimals, and Percents, one parent commented on the newly gained content knowledge and how this knowledge would benefit the child (Knapp, Landers, and Jefferson 2012, p. 30):

Parent A: For example, one night, we had this conversation: A half ... what is the half of a quarter?

Interviewer: Oh?

Parent A: And would you believe that for years I didn't know what half of a quarter ...

Interviewer: Half of a quarter.

Parent A: It is one-eighth.

Interviewer: Yes.

Parent A: And that you keep cutting it [*The fraction strip*] ... umm ... $1/2$ of $1/8$ ...

Interviewer: So ... you know. OK.

Parent A: And even on this test that I got ... they asked me that question ... $1/2$ of a quarter, and I could answer.

This parent learned conceptually that $1/2$ of $1/4$ is $1/8$ through a fraction-strip activity in the minicourse. Program evaluation has shown that participating family members gained content knowledge about turning percentages into fractions, calculating the volume of a cylinder, and raising a nonzero number to the zero power, obtaining 1 as the result. They also learned pedagogical strategies, such as using base-ten blocks or tan-grams to assist their children in developing conceptual understanding related to number and geometry.

Participating teachers commented that they believed it was important for children to see their parents taking risks and persevering with mathematics. In addition, because teachers and family members shared instructional experiences, they were able to discuss the details and nuances of delivering mathematics instruction.

The MAPPS project creates avenues for teachers and parents to learn how to deepen a child's conceptual understanding of mathematics. Taking courses together allows teachers access to parents' multiple funds of knowledge. It also forges productive pathways between families and teachers for continued discussions about mathematics teaching and learning strategies to support a child's mathematics identity, thus transforming family involvement for parents and teachers alike, both of whom have powerful influences on a child's life. One parent described the gains from the program:

> The point is to be part of the school and part of the community, just like a student. For me, the main part [of the MAPPS program] was parental involvement to use other parents to teach parents. I want to be part of that. (Civil and Bernier 2006, p. 328)

Partnering with Community Organizations

In addition to family partnerships, community organizations can serve as instructional resources to support children's mathematical learning. In an attempt to support families, many community and faith-based organizations have established children's educational programs. They exist in local neighborhoods, making them closer and easier for families to use. This is particularly important when students cannot take advantage of before- or after-school supports or enrichment services.

This was the case for students attending Stuart Middle School, a California state-designated "distinguished school," known for its innovative instruction and high levels of student achievement. Stuart, a school in an urban district, is located in an upper-middle-class residential area that has recently experienced a demographic shift. As a result, the school draws an increasing number of students from the racially diverse and working-class "flatlands" communities in the district. Many of the flatlands students must get up at 5:30 in the morning and take several public and district buses to and from school each day.

The mathematics department at Stuart offers after-school tutoring and enrichment activities such as robotics. Unfortunately, if students don't live in the neighborhood or have someone to pick them up from school, they must take the district bus that leaves right after school. Because of bus schedules, the principal and staff had been unable to arrange for the students who live in the flatlands to take advantage of additional supports and enrichment programs. A teacher and a Title I community liaison (identified as Pam and Margaret, respectively) were talking about this problem in the faculty room during a recess break:

Margaret:	Hi, Pam. How are the parent conferences going? I saw your schedule on the door. Boy, you have a packed schedule.
Pam:	Yes, I have four more to go.
Margaret:	What's wrong? Are they not going well?
Pam:	No, they are all right, but I can't seem to get the parents of the kids I need to talk with to come to conferences. I sent several e-mails. Several of these students are way behind the other students in the class. They need more math support. Other students could really benefit from being on the robotics team. I tried to get them to attend the after-school sessions, but they have to catch the 3:15 bus.

Margaret: You know, until I had a child, I really didn't understand how much I was underutilizing a very powerful tool. Parents have busy lives and aren't always able to support their child in the ways schools select. You know, I think many of our parents attend First Congregational Church. The church has a tutorial and after-school center. There's also Taylor Baptist Church. I think I can get the phone number of each of the community directors. I know it's another step, but it is definitely worth the time if we can forge a partnership with these community churches.

In this case, the community liaison was able to meet with each director and set up meetings with the mathematics department chair, the principal, and the parent advisory council. The group decided to pursue a federal grant to provide professional development, enabling church staff members to articulate connections between content offered during the regular school hours and that covered in out-of-school programs. The church staff members would also gain pedagogical content knowledge, allowing them to infuse mathematics strategies into their student educational programs. Margaret also planned to talk with the directors of the local Boys and Girls Clubs to determine whether they would like to join the partnership and perhaps sponsor a second meeting space for the robotics team. In the interim, the members of the group agreed to include one another on all mathematics-related correspondence with families regarding curriculum and assessment issues. Church and community directors were also granted access to the school's electronic newsletter and bulletins to increase community–home–school articulation.

Collaborating with church or community organizations can help teachers address student and family educational needs that the school system cannot effectively tackle on its own. The mission of these community stakeholders is to support children and families so that they do not feel marginalized in their schools. Realizing the connection between mathematics success and a child's high school and college options, many of these stakeholders have directed their efforts to address educational access. Partnering with the faith-based and community resources in a child's life can provide teachers with opportunities to build on and deepen the mathematical development that they are trying to address in their mathematics programs. Furthermore, using the multiple resources in students' communities positively influences how they view themselves as mathematics learners and how they, in turn, are viewed by others in their communities.

Conclusion

A teacher must weave together many threads to create an environment in which each child feels and performs like a capable and confident mathematics student. Unfortunately, it isn't always easy to gather and tie together all the necessary threads. Parents and community groups are an underused resource for increasing students' mathematics proficiency.

Partnering with critical stakeholders can enable a teacher to implement many of the equitable practices stressed in this book, such as *going deep with mathematics, affirming mathematics learners' identities, and drawing on multiple resources of knowledge.* One of the inherent benefits of any partnership is that its members bring different resources to the table, and, by pooling these, all parties can secure what they need to reach their mutual goals. Teachers, parents, caregivers, and community stakeholders clearly have a common goal: increasing students' mathematics success.

DISCUSSION QUESTIONS

1. How would you describe your parents' and caregivers' level of support for their children's mathematics education? What evidence contributes to your description?

2. How would you describe your parents' and caregivers' level of participation in their children's mathematics education in school and out of school? What evidence contributes to your description? What would you like to change or stay the same?

3. What resources at your school are available to families who want additional knowledge or strategies to support their children's mathematics instruction?

4. What community organizations could you partner with, or what community resources could you use, to secure further support for students in your classroom?

Chapter 12

(Re)Connecting with Families about Math Learning in Middle School and High School

This chapter focuses on reimagining the connection with families at the secondary level to promote positive mathematics learning and identity. We will offer three vignettes that illustrate ways schools and families can communicate about mathematics education at the secondary level. As children move from elementary to secondary school, school-based relationships with families transition as well. Families must navigate a new system of classes and schedules. Students have six to eight teachers a day rather than one to three like in elementary schools. Secondary schools are often larger and more crowded. Youth are developmentally changing in mind, body, and spirit. Secondary mathematics teachers are responsible for 120–180 kids per day. Curricular tracking becomes explicit with children being ranked and sorted into particular math tracks often starting in seventh grade. For example, the high school appendix of the Common Core State Standards in mathematics created two pathways starting in seventh grade: compacted track and the common core track. Compacted track leads to high school algebra by eighth grade. Common Core track leads to algebra taken in ninth grade (National Governors Association Center for Best Practices and Council of Chief State School Officers 2010). These tracks or pathways often determine access to college and career opportunities. In addition, teenagers are often studying domains of mathematics (e.g., algebra, geometry, proportional reasoning, probability) that many adults, including parents and caregivers, have not taken nor studied deeply for a long time. Helping teenagers with homework is not as easy as it might have been with younger children in elementary school.

Research shows that family engagement is critical to positive student learning outcomes (Ishimaru 2019). In fact, through case study research we know that Black and Latinx parents often must advocate for their children to be given opportunities to pursue college-bound coursework because of policies in place designed to maintain systemic inequities such as tracking that disenfranchise children and communities of color (Adams Corral 2019; Berry 2008; Martin 2000, 2006). However, there is limited research about what positive family engagement looks like at the secondary level, especially about mathematics education.

The goals of this chapter are to critically examine communication between schools and families about mathematics education and discuss family engagement strategies that promote communication, collaboration, and partnership to help young people and their families feel empowered to make informed decisions related to mathematics education. We will start with a vignette to illustrate the complexities and commitments that affect family engagement and youth experiences with mathematics in the secondary setting. This vignette focuses on an e-mail exchange between a parent and an eighth-grade algebra teacher. Individual e-mails as well as group e-mails sent through the academic system platforms are often the primary ways teachers communicate with families. The second vignette explores a systems example of communication about math education to families. In this case, through a district communication, sixth-grade families are asked to select a math pathway for their child. We will focus on how issues of power and identity are implicated and consequential to relationships with families, particularly families from groups whose knowledge and experiences have been historically marginalized in education. The third vignette offers another approach to communicate with families about changes in math programs that could significantly influence their child's experience with mathematics and access to advanced math courses. All three vignettes provide space for critical analysis about how school-based relationships with families play a critical role in access to and advancement in mathematics.

Vignette 1: Reaching Out for Help

This vignette is based on a story Julia Aguirre has shared with many mathematics educators at various speaking engagements since 2016. The story resonates for different reasons with a variety of audiences including parents, caregivers, teachers, coaches, and administrators. This vignette centers on a mid-December e-mail exchange between a parent and the eighth-grade algebra 1 teacher after a chapter test. For some context, this middle school followed the Common Core Standards for Mathematics (National Governors Association Center for Best Practices and Council of Chief State School Officers 2010). This included the mathematics tracks set up by the Common Core High School (appendix A, pp. 80–81) that outlined specific tracks to reach algebra by eighth grade (compacted track) or algebra by ninth grade (Common Core track). The middle school had both tracks. All children took sixth-grade Common Core mathematics and then split into different tracks starting in seventh grade. We will start with the parent's e-mail communication.

> Dear Eighth-Grade Algebra Teacher,
>
> Do you have any recommendations for tutors in the local area? Our daughter, Ana, is struggling and we need a third party to help. Are there peer-tutors available at the middle school? We are happy to bring her to get help from you, but we also need additional resources. Ideas would be helpful. We are trying to

support a growth mindset. But this is a challenge when she continues to get Fs on her tests. Very discouraging for her as you can imagine.

—Parent

STOP AND REFLECT:

How might you respond to this parent's e-mail?

What stands out to you in relation to Ana's math identity and experiences?

Why is this parent reaching out? What is the parent worried about? What is the request?

The parent has specifically asked for tutoring resources at the school or local area. The parent recognizes that their child is struggling with learning the content and does not feel equipped to personally help in this situation (e.g., "need a third party," a "tutor"). The parent is concerned about their child's math identity and confidence, given consistently low grades (Fs) on tests.

The following e-mail excerpt is the response the parent received from the mathematics teacher. The response is lengthy and detailed, but as you read it, take note of what the teacher emphasized in the response and what type of message the parent might receive about Ana.

Hi Parent,

Thanks for your message. I did ask our math department chair about any tutors she is aware of, and she mentioned this website www._____. I don't know of other specific tutors at this point. Certainly there are teachers at Math Lab on Thursdays who are happy to help (and sometimes some high school students come as well, though not always). I spent some time looking at Ana's math history, and I noticed that her state test scores are lower than many of her peers on the compacted path (typically at a Level 4).

Third grade—Math Level 2

Fourth grade—Math Level 1

Fifth grade—Math Level 2 (a Level 4 in reading and science though that year—that's awesome!)

Sixth grade—Math Level 2 (grade in math fluctuated between an A and a D+—ended the year with an A)

Seventh grade—Math Level 2 (grade in CC 7/8 fluctuated between a B and C—most frequently a B−, ended the year with a C)

Certainly, I understand that those scores are not measuring all of her knowledge and understanding, but at the same time, I notice that she often seems to be struggling to keep up.

Her grades in CC 7/8 also suggest that she only had a partial understanding of many topics from last year, which makes it difficult to build on that understanding this year. As I noted in another message, the curriculum is going to get more challenging at this point, because the earlier units were much more of a review from last year and we are now doing a lot of content that is very new to students (such as operations with radicals). After our unit on the Pythagorean theorem we proceed into exponents and exponential models, functions, transformations, and quadratics (we spend a significant amount of time on quadratics in particular). I think it's great that you want to get Ana more support, though I also think it's worth noting that there is another option you may want to consider. Sometimes we have students in algebra who are having trouble keeping up and we have them change to take CC 8 so that they can take algebra next year in high school. Students in CC 7/8 did most (but not all) of the CC 8 curriculum, so there are concepts that they benefit from going through in more depth to be prepared for high school algebra. I know Ana is a very intelligent, capable student, though as I've mentioned previously, I notice that she often needs more time to process through the concepts that she's learning and she gets distracted easily (so she doesn't always maximize her class time, which causes her to miss out on getting more feedback and support). I am happy to have her as a student, though I just wanted to mention that CC 8 is an option if algebra is feeling too overwhelming. We want students to earn a grade in algebra that they want to apply to their high school transcript (a high grade that will help their GPA) and to feel prepared for geometry. Please feel free to let me know if you have other questions or concerns or would like to speak about this more (in person or via phone). Hopefully having a retake tomorrow on the Pythagorean theorem quiz and a retake on Friday for the Chapter 5 test will help her grade significantly. Ana has her test and her quiz with her that she can use to review tonight and tomorrow as needed.

Take care,

Teacher

STOP AND REFLECT:

What lens is the teacher using to respond to the parent about what's going on with Ana?

What math learning and identity issues are raised for the student, parent, and teacher in this situation? What role do assessments and standards play in this situation?

The thorough response by the algebra teacher reflects a commitment to wanting to help in this situation. Although only able to provide some general resources related to tutoring, the response centers on providing the parent alternatives to address this child's current struggle with mathematics. The teacher's initial response was to see if there was a history of problems with mathematics. The teacher chose to review the student's state test score data (grades 3–7) and letter grades (for math courses taken in sixth and seventh grades). The teacher framed Ana's past performance compared with her class-mates in the eighth-grade algebra class. According to the teacher, the students in the eighth-grade algebra class consistently scored at the highest level (level 4) on state tests throughout elementary and middle school. Ana's test scores in mathematics did not reach that level at any time in grades 3–7. Although she did achieve a final letter grade of A in sixth-grade Common Core math class, the teacher noted grade fluctuations across both sixth- and seventh-grade math classes, which the teacher suggests contribute to partial understanding of math concepts. Ana's performance in 7/8 compacted track was a level 2 (out of 4 which is approaching standard) state test score and a final letter grade of C. The pattern of low performance in mathematics as measured by test scores and fluctuating grades in middle school are significant pieces of evidence the teacher felt the parent should know that might explain the current struggles. Ana might have only partial understanding of the material, causing challenges in the algebra course. The teacher also felt compelled to let the parent know that most of the content covered in the algebra class so far was a review from the previous year (CC 7/8). Thus, the upcom-ing content would be new to everyone, and the teacher predicted Ana might fall further behind since she is currently struggling with material she should already know. What is clear from this first part of the response is that test scores really matter to this teacher as a strong predictor of eighth-grade algebra performance. Although research has shown that reliance on test scores as a sole measure of mathematics proficiency is highly prob-lematic, it is all too often used in this manner to reflect current and future math learning (Berry, Ellis, and Hughes 2014; Davis and Martin 2018). Furthermore, this emphasis on test scores puts the responsibility of performance on the individual child. Quality of instruction and access to academic enrichment resources (such as private tutoring) are not considerations.

The second part of the response focuses on the concern that Ana will have diffi-culty keeping pace with the current class. The teacher suggests that the parent consider transferring Ana out of the algebra class to the lower-track class (eighth-grade Common Core math). The teacher recognizes this as a viable alternative for students who may "have trouble keeping up" or find the class "too overwhelming." Although the teacher acknowledges Ana's intelligence, the teacher also shares with the parent that Ana is distracted easily and "misses out" on extra support and feedback provided in class. Ana continues to be the source of the learning problem. The solution offered by the teacher is to transfer the child out of the class.

The response ends on a hopeful note that with retakes on a recent quiz and test, Ana's grade might improve. The teacher has a revision policy that offers opportunities to improve grades. However, it is unclear what instructional support, other than studying the previous assessments, is or has been available to help prepare for the retake.

This exchange reveals some of the complexities and tensions teachers, families, and children face in school mathematics. Consider the following questions:

- How is mathematics learning defined in this exchange?
- How does this exchange reflect views about the child's mathematics identity?
- In what ways are standards affecting this situation?
- How does this exchange highlight tracking and ability grouping?

- What if the parent follows the teacher's advice and switches the child to the lower track? What assumptions about the mathematical experience are being made?

- How does this exchange reflect parent-teacher relationships?

- If you were the instructional coach, how might you advise this teacher to address this situation?

Answers to these questions should raise significant concerns among mathematics teachers, coaches, and leaders. Applying a critical lens would encourage examining systemic policies and practices that affect teaching, learning, and family engagement. For example, how are students placed in specific tracks (placement policies and practices)? What assessments are used to make high-stakes decisions (assessment policies and practices); are test scores used as the only predictor or the priority predictor of performance? Is that problematic? And how are families made aware of these practices and policies? What are the power dynamics related to engaging families or not about school placement and assessment policies?

In addition to the systemic concerns, this situation raises social-emotional concerns. How does a student's identity change if that student is moved to a lower track? How does this affect the student's agency? How does this affect the student's social connections (i.e., friendships)? What message does it send to parents and caregivers about their child? After a recent presentation, a parent shared with Julia Aguirre that a similar situation happened to her middle school daughter, and the daughter tried to end her life. The consequences are real. There *must* be better ways to partner with families and students to support mathematical joy, access, and advancement, especially if the road is difficult at times.

Vignette 2: Choosing Pathways

Current efforts to create more math pathways or "branches" for student advancement open up opportunities for communication with families and students to make informed decisions about mathematics education. Yet, there is limited research to help guide this work. It is important to think about how the purposes and goals of mathematics education are communicated to families. For example, how do families come to understand the mathematical pathways available to students and the consequences of those pathways on students' future life opportunities and aspirations? How does the school work with families to pick math classes? The following is an example of how one district communicated with families about mathematics pathways. As you read, note instances of power dynamics and voice at play that might influence how students ultimately end up in specific courses.

Notifying families about middle school math classes

In the spring, a three-page letter from the district superintendent's office was sent home to families of sixth graders. The letter, written in English, focused on the math course selection for the following year. There were two choices: Common Core grade-level pacing, which would lead to algebra 1 by ninth grade, and the compacted-track grade-level pacing, which would lead to algebra 1 in eighth grade. Algebra 1, in this case, was the first high school credit-bearing class that would count toward graduation and college admissions. Previously, this district adopted an eighth-grade algebra 1 mandate, meaning all students would take algebra 1 by the eighth-grade. However, the district's shift to the Common Core standards now provided a choice for families.

The purpose of the letter was to "explain placement options for current sixth graders." The letter provided some background information of how these middle school courses were developed, stating they were "based on recommendations in the Appendix A: Designing High School Mathematics Course Based on the Common Core State Standards for Mathematics (CCSS-M)." This was followed by a diagram that showed the different pathways through twelfth grade (see figure 12.1). There were asterisked statements to remind families that students needed to take three years of mathematics in grades 9 to 12 to graduate from high school. And that state tests would be given in sixth, seventh, eighth, and eleventh grades. Passing the state test was also a graduation requirement.

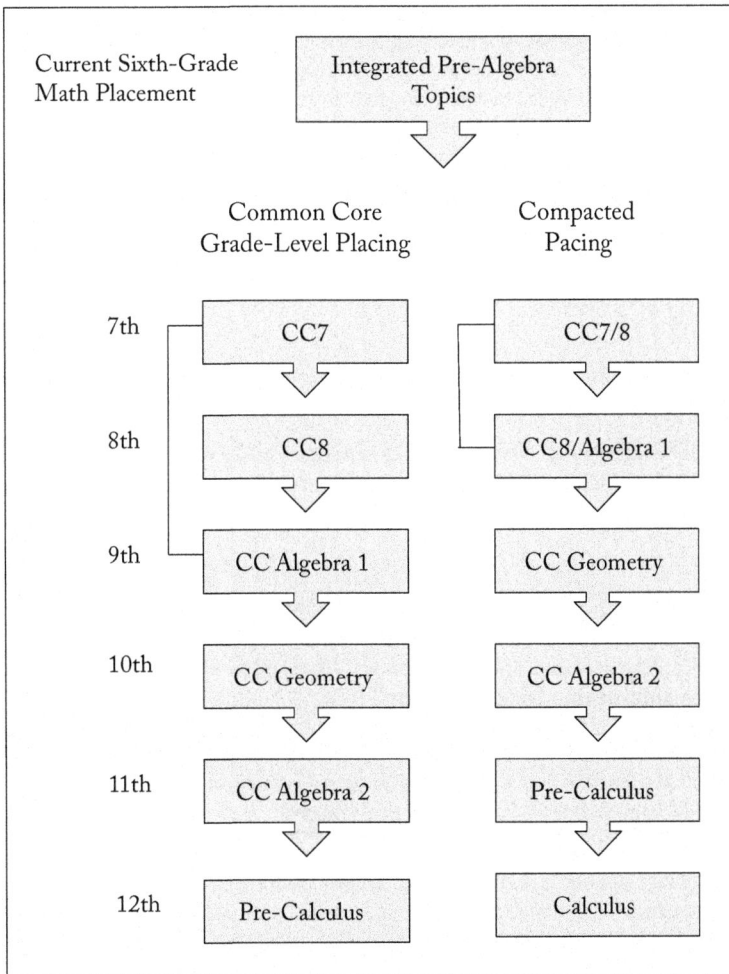

Fig. 12.1. Flowchart diagram depicting secondary math pathways

The letter then provides a section called "What's Different." In this section, there are descriptions about Common Core math courses, the compacted courses, and the standards of mathematical practices (see table 12.1).

Table 12.1. "What's Different" descriptions

CC7 and **CC8** are grade-level courses aligned with CCSS-M and are more focused, coherent, and rigorous than our current courses. **CC7** and **CC8** will build a strong mathematical foundation for students in middle school.
CC7/8 and **CC8/algebra 1** are for students who choose faster pacing. • Three years of math standards (**CC7**, **CC8**, and **CC algebra 1**) are compacted into two years of instruction. • This is accomplished by dividing the eight-grade standards between two courses. • **CC7/8** consists of all of **CC7** plus a portion of the eighth-grade standards. • **CC8/algebra 1** contains the remaining eighth-grade standards plus the full content of CC algebra
STANDARDS FOR MATHEMATICAL PRACTICE: Common Core expects much more than just procedural skill and fluency. Students are expected to achieve deep conceptual understanding, to demonstrate critical thinking skills, to make connections, to persevere in problem solving, to apply their learning to solving problems in real-world contexts, to justify their reasoning, and to clearly communicate their math thinking. To read more about the Standards of Mathematical Practice at http://www.corestandards.org/Math/Practice.

This was followed by a much smaller section called "What's the Same?" In this section, the following bullet points were offered:

- Families can still self-select the most appropriate math pacing for their child.
- All students will have the opportunity to study linear functions in middle school. Much of the content central to typical algebra 1 courses—linear equations, inequalities, and functions—is found in CC8.

The third page of the letter offers "considerations" to help parents and caregivers decide, including "readiness criteria" for compact pacing (see table 12.2).

Table 12.2. Considerations for placement

COMMON CORE GRADE-LEVEL PACING: Students are expected to master procedural skill and fluency, conceptual understanding, and application to real-world problem solving. Students are also expected to achieve the Standards for Mathematical Practice.
COMPACTED PACING: Students who select compacted pacing have the same expectations for Common Core grade-level pacing with respect to achieving content standards and standards for mathematical practice. No content will be skipped, so, with faster pacing, students will complete three years of Common Core mathematics in two years of instruction. Therefore, students who select compacted pacing should be prepared for increased rigor and vigorous pacing. Students who will be most successful with compacted pacing exhibit strong organizational skills, maturity, self-motivation, and affinity for mathematics.

Table 12.2. Considerations for placement (*Continued*)

HOW DOES THE PACING DECISION AFFECT HIGH SCHOOL? There will no longer be a self-select decision at the end of eighth grade. Students who select Common Core grade-level pacing will begin high school in CC algebra 1. Students who select compacted pacing will begin high school in CC geometry. <u>Our recommendation is to select the middle school math pacing that best meets the student's current level of readiness</u>. There are many possible sequences of courses through high school offering opportunities to accommodate academic goals in mathematics.
READINESS CRITERIA: Families who wish to select compacted pacing should ensure that their student meets the following readiness criteria: • An average of 90 percent or higher in the assessment category of their current sixth-grade math course (Students who consistently earn As on tests and quizzes indicate readiness for faster pacing.) • A percentile of 95 or higher on the fifth-grade Stanford Test • A score of 475 or higher on the fifth-grade State Test • Consistently demonstrates Standards for Mathematical Practices

This table was followed by a readiness checklist to select compact pacing, in which parents and caregivers had to provide evidence of test scores and assessment grades and place their initials that confirmed the statement:

My child consistently demonstrated the standards for mathematical practice by achieving deep conceptual understanding, demonstrating critical thinking skills, making connections, persevering in problem solving, applying learning to solving real-world problems, justifying reasoning, and clearly communicating math thinking.

A signature form that noted which option plus the readiness checklist were to be returned to the school by the child in their current sixth-grade mathematics class.

STOP AND REFLECT:

What stands out to you about this district's communication to families?

What message does it send about mathematics?

What appears to be valued about learning mathematics?

What assumptions do you think are being made about parents' and caregiver's abilities to make sense of this information?

Does this communication privilege some families over others?

This communication reflects an attempt to help families make informed decisions about which mathematics pathway would be best for their children. Families retain the right to self-select the most appropriate math pacing for their children. They are also notified that this self-selection process is happening at the end of sixth grade rather than at the end of eighth grade, which was the previous policy because all eighth graders took algebra 1. Furthermore, there is a bolded and underlined statement giving the district's recommendation that families need to make their choice on the basis of "current level of math readiness."

What is particularly notable is the list of considerations offered by the district. Although there is an explicit attempt to call out linear functions, equations, and inequalities as math content that would be addressed in either pathway, there are specific framings about the rigor, pacing, and type of student that best fits with the compacted track. In the compacted track, the words *faster pacing* and *vigorous pacing* are used. There is also *increased rigor* associated with the compacted track even though there is overlapping content between the two pathways. What is particularly problematic is the sentence summarizing specific student characteristics of those who will be successful in the compacted pacing track: "Students who will be most successful with compacted pacing exhibit strong organizational skills, maturity, self-motivation, and affinity for mathematics." This sentence suggests that those placed in the alternate track do not have these characteristics. They lack maturity, self-motivation, and organizational skills. Moreover, there needs to be an affinity for mathematics to be in the compacted track. This suggests that a child needs to show a natural liking for or inclination toward mathematics to pursue this option.

The emphasis on "readiness," "maturity," and "affinity" in this communication reflects a fixed mindset that has far-reaching implications. The readiness criteria rest on achievement levels of standardized tests and curriculum-based measures. Only top scores are identified, and these top scores are for the child's previous year's mathematics performance. Parents are being asked to use the child's fifth-grade test scores to select a seventh-grade math pathway that will affect when that child takes algebra 1 and ultimately whether the child will go to calculus or precalculus.

It is hard to imagine how a parent or caregiver might feel trying to navigate this important document. The document is only available in English. There is no contact person to ask questions. There appears to be an assumption that families are familiar with the Common Core standards, affiliated math content, and the role of standardized tests in this process. The wording of the documents puts the primary responsibility on families rather than educators with sentences such as "Families who wish to select compacted pacing should *ensure* [italics added] that their student meets the following readiness criteria." How can families ensure that their child consistently demonstrates the standards of mathematical practice? What is the responsibility of the school and teachers to ensure that students meet the readiness criteria? In the end, the communication to families suggests that math readiness is predictable, fixed, and the responsibility of the family and student, not the school, math department, or teachers.

What might be a more holistic family-centered communication about math pathways for secondary students? Next, we will explore an alternative approach to working with families in secondary math settings.

Rethinking Family Partnerships for Mathematical Advancement

Current initiatives such as the Branching Out Report (Daro and Asturias 2019) and the Launch Years Report (Dana Center 2020) reframe the math pathways discussion to include more opportunities to take more mathematics for college and career readiness. The goal is to offer advanced math courses such as data science, statistics, and mathematical modeling as alternatives to precalculus or calculus courses. Emerging research shows that such courses can have a positive impact on students' sense of belonging, agency, and competence (Heinzman 2022). In one study, students taking an Introduction to Data Science class had increased engagement in the course. In contrast to previous math experiences including failures, students in the Introduction to Data Science class perceived mathematics as more meaningful, and even "cool" (Heinzman 2022).

However, preliminary research also suggests that despite best efforts, sometimes "branches" are reinforcing tracking with negative impacts on math identity (Smith Arrillaga et al. 2023). For example, there is a perception that those students who choose an alternative pathway to precalculus are taking less rigorous courses than those pursuing precalculus and then calculus. This reinforces status labels that suggest precalculus track is for the smarter kids. If demographic trends hold, then precalculus will remain mostly white, Asian, and middle class, and the other tracks will serve students from working-class communities and will be more racially and ethnically diverse.

These new pathways reports emphasize attention on school guidance counselors to support student agency to make decisions about mathematics course work. The emphasis is on empowering young people to make informed decisions about their classes that will help them achieve their career and college aspirations. Although the focus on guidance counselors is important, it is not enough. According to the American School Counseling Association, the national average of students to counselors for 2021–2022 was 408 to 1 (2023). The American School Counseling Association recommendation is 250 to 1. More importantly, underresourced public schools serving students of color and working-class students are more likely to have higher student-to-counselor ratios than well-resourced schools. A focus on strengthening guidance counselors' understanding of mathematics education is a limited strategy since student access to those counselors is scarce. We suggest multipronged approaches that center on family and community partnerships to help students make informed decisions about what math pathways to pursue while keeping options open as students nurture talents, develop new interests, or have positive enrichment experiences that offer glimpses into various careers, including STEM careers. There is limited research on such partnerships in middle and high school settings. We must ask ourselves why and how we can do better because it is the families and caregivers that connect to young people daily and perhaps know them best. We encourage schools to refocus their partnerships with families and communities.

Some caveats are needed to help clarify our perspective on communicating with families about math pathways. First, we believe that all students should take meaningful and relevant mathematics courses throughout middle school and high school that support mathematical, social, and knowledge empowerment. This makes the branching pathways starting as late as possible, such as junior year of high school, appealing. However, we must be diligent that all the math courses students take lead to experiences that foster curiosity, collaboration, and connection rather than competition and "vigorous pacing," whether this is data science, mathematical modeling, or precalculus. As a math community, we must also work hard to break down course stereotypes reflected in vignette #2 that suggest that a student's "readiness," "maturity," and "mathematics affinity" predicts performance. These attributions are relative and vulnerable to implicit and explicit bias. Returning to Chapter 4 and Chapter 5, are these standards-based courses and assessment practices worthy of our students? All these courses need to be worthy of our students, which means that academic apartheid (i.e., tracking) must be eliminated to provide students with true choices (Wells 2018).

Finally, we must look outside of the mathematics classroom to informal or nontraditional learning spaces such as enrichment programs, culture-based clubs and student groups, or extracurricular activities occurring on school grounds and in the community as places students may be thriving. This would enable us to gain valuable insights to help re-create some of the conditions that have shown positive impacts on student identity and learning.

Vignette 3: Families on Board—Demystifying Detracking through Multilayered Approach: Aldercreek's Story

In 2019, prior to the pandemic lockdown, math teacher leaders at Aldercreek Middle School in Oregon decided it was time to detrack. Aldercreek is located in an urban region of the state with a growing Latinx student population of 30 percent. The demographics of the school included 24 percent designated as multilingual learners, and 95 percent of the children were affected by poverty. The decision to detrack was mainly due to the use of a standardized math placement test given at the end of sixth grade that placed students into the math 7 track or compacted math 7/8 track. One teacher, Ms. T, noticed that many of her seventh graders blossomed as mathematicians, but because of the standardized test scores, they were not in the higher track (compacted 7/8 track). In addition, another teacher, Ms. M, noted that the demographics of the algebra 1 course (eighth grade) did not match the demographics of the school, specifically for Latinx student representation. Because of these observations, Ms. M and Ms. T began to research schools and districts that attempted to detrack math courses. They were closely watching the work being done in the San Francisco Unified School District (Torres, Nguyen, Barnes, Wentworth 2023). They also knew that if the detracking initiative was to be successful, a broad base of family support was critical. It is important to note that math reform efforts often engage families from dominant cultural groups, white and middle class, through traditional activities such as family nights, PTA meetings, and booster clubs. Research has shown that families from historically marginalized and segregated communities are often not included in these conversations (Barajas-López and Ishimaru 2020; Barajas-López and Larnell 2019; Ishimaru, Barajas-López, and Bang 2015). Ms. M and Ms. T wanted to make sure that outreach work with families about detracking plans was intentionally inclusive. Ms. M explains:

> So we needed to get to where they were already congregating. That was a high priority for us. And so, in the span of two or three months, we probably hit as many clubs that were already meeting. We have a dual language program that has a Spanish-speaking family night that we got on the agenda for. We were launching a Black student union at the time, so we made sure that we got to some face time there.

Their goals were to explain the new math pathway the middle school would be implementing, help families understand what that meant for their children's learning experience, and dispel any concerns families may have about their child's access to more advanced math courses and college.

In addition, outreach to families also included repurposing a family night that coincided with an orchestra concert. They took this opportunity to address concerns by some parents who wanted their children in the 7/8 compacted track. Ms. M and Ms. T provided an overview of what a typical math lesson would look like in the Common Core Math 7 pathway. The parents responded positively. Meeting with this group of parents preemptively was an important strategic move to prevent preconceived notions from forming that might oppose this change.

Six months were devoted to family outreach about this change. Then, COVID-19 hit. Despite this uncertain scary time, the initiative did not stop. And, in fact, it may have created opportunities to accelerate the process.

A second approach to bring families on board with this decision was to recognize that families engage with more school staff than just math teachers. Other staff members, including counselors and other teachers, needed to be able to answer questions about what was happening with the math courses. As Ms. M explained,

We also knew that math teachers were not going to be the only touch points that families were going to engage with. And so, we developed a one-pager that went out to our whole staff with bullet points of the research behind what we were going for. It went to counselors as well.

This one-pager was then further developed into a district-wide public awareness campaign to alert families what was going on at the middle schools district-wide. Face-to-face meetings were not possible because of COVID-19, so a letter from the sixth-grade principals along with a one-page resource flier (see figure 12.2) was sent to parents and caregivers of middle school children to attend online focus group meetings to get parent feedback. It was translated into four languages (Spanish, Russian, Vietnamese, and Chinese).

This flier contains an abundance of information and live links for families to further deepen their knowledge about the proposed changes in the math program. There are several multimodel venues to better understand research, policy, and practice considerations of detracking. For example, the Math Considerations link takes you to a one-page description that explains the tracking context of the district leading to "disparities" in access to college preparatory courses. It also lays out the beliefs and proposed structures needed to support this shift (see figure 12.3).

The content of the math considerations document is explicit that calculus is one of many college preparatory math courses. This is key to keeping the empowerment door open as wide as possible—mathematical, social, and knowledge empowerment for students. It also explicitly addresses status labels attached to specific courses, who have traditionally had access to those courses, and affirms there are many ways to be "smart" in math class. The second set of bullets provides examples that send a message to parents and caregivers that their child will be appropriately supported in the rigorous standards-based math courses. Rigor and support are key.

NCSD 6-8 MATHEMATICS:
Español Русский Việt 中文

Reviewing and Revising NCSD's Middle School Math Program

Math Focus Group Goal: Create a rigorous middle school math program to establish a foundation of deep mathematical understanding in order to prepare each student for multiple pathways to advanced level coursework in high school math and beyond.

Research

Articles about math education
- Forbes, Dec. 2020
- USA Today, Feb. 2020
- Edutopia, Aug, 2019
- NCTM, March 2018

Research & Position Papers
- YouCubed Study
- Student Experience Network Study
- National Council of Teachers of Mathematics
- National Education Policy Center

Case Study: San Francisco
- San Francisco USD Overview
- Quick Facts about Math and Tracking

NCSD Work

June 2021 Math Update Video Overview

Math Considerations

Rationale and Frequently Asked Questions

Oregon Department of Education Information
- Presentation to NCSD teachers (video)
- ODE Mathematics Website

Instructional Shifts

Videos
- Five Principles of Extraordinary Math Teaching
- Rethinking Giftedness
- Dr. Jo Boaler on Ability Grouping
- Understanding "Old" and "New" Mathematics
- Rich Tasks: Square Counting video

Sample Math Lessons
- Number Talk
- Using a Worked Example
- Multiple Ways to Solve a Problem

Notes:
- Any changes to our Middle School Math Program will necessitate changes to Elementary, High School, TAG, and SPED programs. These changes are being considered and information will be shared as we move forward.

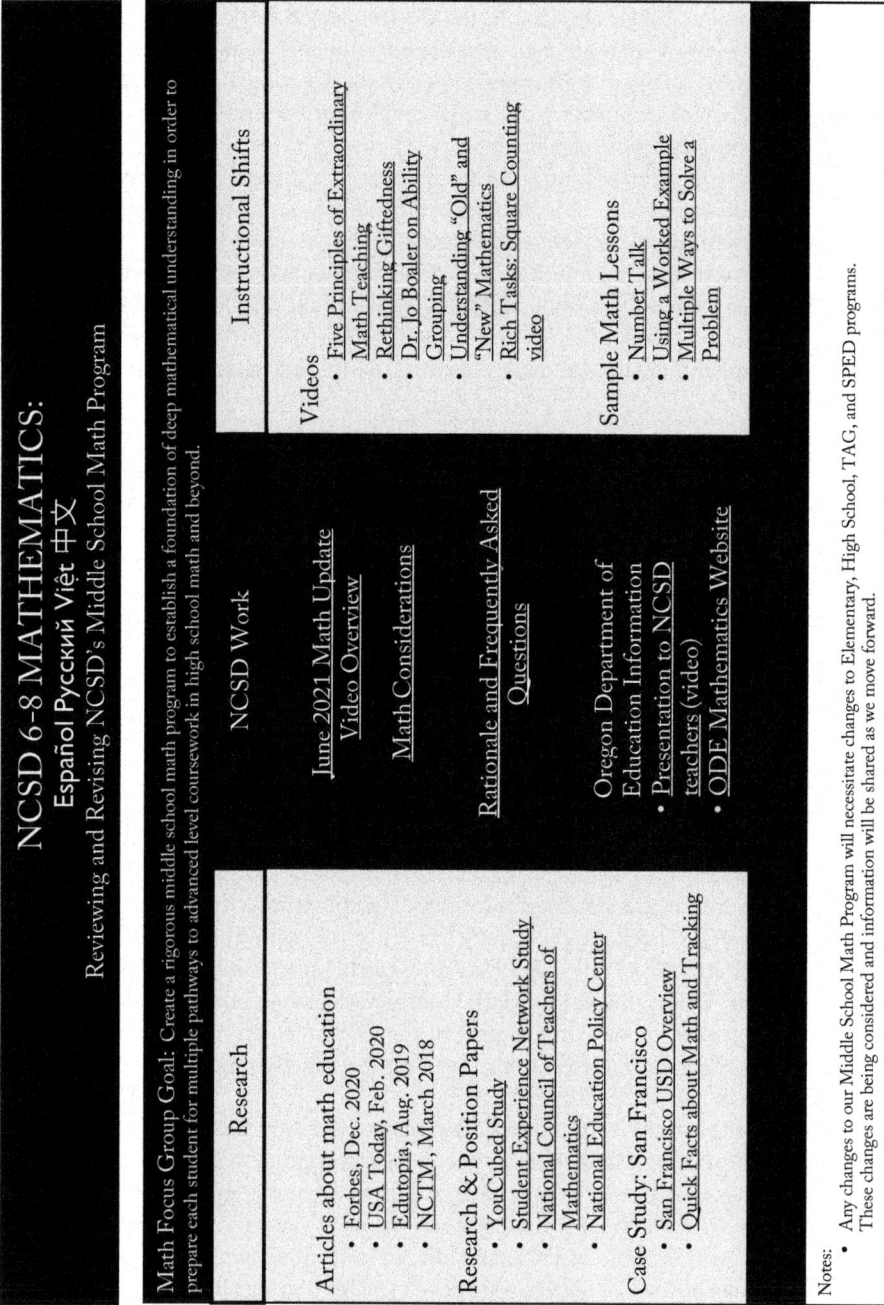

Fig. 12.2. District information flier

We believe ...

- All students deserve access to a rigorous, high-quality math education at all grade levels.
- There should be multiple pathways to calculus or other college-preparatory math courses.
- Middle school Common Core standards are foundational for success in all future math courses.
- All students should have access to college-preparatory math courses, not only those who are labeled "advanced" after sixth grade.
- Students learn best in mixed-ability classes that celebrate multiple ways of being smart in math.

To this end, we are proposing modifying the math courses in middle school to ...

- Offer common courses for all students with appropriate supports and extensions embedded for each student
- Offer ample opportunities for students to master foundational topics in mathematics before moving on to high school math
- Focus on conceptual understanding and connections in mathematics, not only on procedural fluency
- Include algebra 1 topics in eighth-grade math, aligned with Common Core and new ODE standards

Fig. 12.3. Public-facing resource to explain district efforts

Along with the supplemental materials, this flier was accompanied by strategic outreach to family groups in the district. For example, Ms. K, the district's math specialist, reached out to Las Madres—a Spanish-speaking group of Latina mothers—to discuss the information. Ms. K practiced the presentation by recording for the group's translator prior to the meeting to address any technical terms that might need additional clarification. The video was then used as an additional resource for the flier that could be translated into multiple languages to reach more families. Families could view the recording on YouTube. Ms. K explained the key message trying to be conveyed:

"We're not taking away advanced classes from your kids. We're giving all kids access to rich tasks all the time so that all kids can take whatever classes they want later."

Ms. K, Ms. M, and Ms. T did share that there was some parental pushback. The survey feedback that was obtained was limited to under ten parents, a majority of whom identified as white with children in advanced classes. Their main concern in the feedback was that access to advanced classes was threatened ("My kid won't be able to take calculus"). This was not the case and was addressed in the resources provided. However, Ms. K's encounter with Las Madres group offered a different perspective on this opportunity.

When I went to the Madres meeting, the parents were very excited about the idea that all kids got access to this. There was a parent who said, "You know, no one ever gave me a choice. I just was in the class they put me in and I'm glad that my student gets to have this opportunity just like everyone else."

Las Madres parents and caregivers recognized this opportunity and realized their own math experience was compromised by tracking. They did not want this for their children. The district listened and moved forward with the detracking plan for the middle school math program.

Aldercreek's detracking journey is still unfolding. COVID-19 affected the family outreach plan, but with intentional planning, families that often were not at the table were identified as crucial stakeholders in the information campaign about detracking. The teacher leaders understood that galvanizing family support for detracking could not be solely focused on convincing the privileged parents. This past approach did not work. For this to work, outreach and support must be broad-based and particularly focused on families with the greatest potential to benefit from this change.

Conclusion

With limited research on family engagement in secondary math settings, we must create new opportunities to center families in the math education journey of their children. This chapter provides examples of how districts, schools, and individual teachers have reached out to parents to address concerns related to math preparation and advancement. These interactions reflect deeper beliefs and structures that influence students' mathematics identity development and learning experiences with mathematics. For some schools and districts, standardized test performance is the determining factor of students' mathematical competency. The message to parents and caregivers is that math learning relies on test performance. This is a fixed mindset approach fueling tracking policies that have long-term negative implications on student identities and experiences. But other schools and districts recognize that traditional ways to measure performance have led to math learning disparities and differential access to college preparatory courses. This needs to change if the goal is equitable mathematics education. Families must be positioned as active partners in this effort to cultivate mathematical agency, collaboration, and more math advancement opportunities for children throughout high school.

DISCUSSION QUESTIONS

1. How do you help families make informed decisions about which math courses to take?

2. What resonated with you in the vignettes? What stretched your thinking? What concerns did they raise for you?

3. How are families centered in secondary math policies and practices in your school or district, especially families from historically marginalized and segregated communities?

4. What practices do you suggest for helping families and their children make successful and empowering transitions from elementary to secondary mathematics?

Epilogue

Enacting Equity and Liberation in Mathematics Education

We hope that this expanded edition book has intensified your commitment to helping children become powerful learners and doers of mathematics. In the first book, we asked you to walk the path toward equity. We also hoped that the messages from the book strengthened your conviction that developing positive mathematics identities is an important—and achievable—goal. Our primary message across both editions continues to be that empowering students in mathematics requires deep reflection on the role of mathematics in students' lives. One critical aspect of your deeper reflection and developing equity-based practice is learning about the backgrounds and multiple identities of your students and how those identities emerge in school and nonschool contexts and learning about your own influence on those identities and how those identities influence the variety of ways in which students engage with mathematics. These identities should not be reduced to stereotypes.

Beyond the critical reflection we encourage you to take action:

- Acknowledge and honor the ordinary brilliance children bring to learning mathematics.

- Dismantle academic apartheid that results from tracking and high-stakes assessments.

- Question seemingly neutral institutional policies and practices, such as standards, grading practices, and placement policies, and how they influence student identity, agency, and empowerment.

- Cultivate authentic partnerships with families and communities that acknowledge the assets they bring to math education conversations.

We have also invited you to reflect on your own personal and professional identities to understand the impact of those experiences on your own identity in relation to mathematics and your instructional practices. As you reflect, you may discover or rediscover that your own experiences as a mathematics learner have shaped your interactions with students and mathematics content. Moreover, you may confirm that, like your students, your views of yourself as a learner and doer of mathematics—your math identity—intersect with other important identities that you have developed.

The five equity-based practices discussed in this book—going deep with mathematics, leveraging multiple mathematical competencies, affirming mathematics learners' identities, challenging the spaces of marginality, and drawing on multiple resources of knowledge—highlight ways that you can strengthen students' mathematics learning identities and agency. You may find that assessing your own teaching in relation to these practices will be a positive first step, or you may have already incorporated aspects of these practices. Using the conceptual tools and practical examples provided in this book may help you to revisit, rethink, recommit, and take action to further empower you and your students.

We acknowledge that changing one's practice takes time. Yet, the need to do so has an urgency that must be taken seriously if we want to see growth in children's mathematical development—particularly among those children who have historically had the least access to and success in mathematics.

Another critical avenue for engaging in equity-based practices is through assessment. Reflecting on the ways in which you communicate math progress to your students and, by extension, their parents and caregivers is key. The models of meaningful feedback provided in this book offer ways to focus students' attention on making sense of mathematics at their current levels of understanding. Traditional feedback, offered to students through check marks, tallies of correct answers, or smiley faces, communicates little to them about how to improve. Grading is a routine practice; rethinking how you give feedback and provide opportunities for revision can strengthen learning and affirm a positive mathematics identity. These are equity-based and humanizing assessment practices that take time but are worth the investment to help students feel empowered mathematically and socially and as knowledge creators.

Because you are not alone in your support of children's mathematics learning, we have also asked you to reflect on the kinds of parent, caregiver, family, and community engagement practices that you use to support mathematics learning and teaching in your classroom. Although routine practices like classroom newsletters and curriculum night events provide opportunities for you to convey your math vision to parents and caregivers, we invite you to utilize a variety of ways to allow all parents and caregivers to share their valuable contributions and insights about their child's math education.

We hope that you share our position that parents, caregivers, and communities can be important resources for supporting children's mathematics learning K–12. Communicating with families and inviting them to be authentic partners can be done in elementary through high school. FAMILY MATH and Math and Parent Partners (MAPPS), highlighted in Chapter 11, are just two examples of programs that support children and families in mathematics. We encourage you to learn about these and similar programs. Furthermore, partnering with faith-based and community organizations can also foster positive partnerships that support children's mathematical development.

Critical reflection on the various ways in which mathematics learning, teaching, and identity are interlinked takes commitment and time. Questioning systemic policies and taking action to change one's instructional practice to accommodate this new knowledge requires courage and perseverance. Changing habits of thinking and pedagogical frameworks may seem artificial or cumbersome at first, but the new practices gradually become easier and more natural. What every teacher wants is to engage and support students to be creative, confident, and competent mathematical problem solvers—*Math Strong*. What we have tried to outline in this book are strategies and perspectives that support this goal and are based on research and our collective experiences as mathematics educators.

It has been our aim to make these equity-based practices more transparent by offering examples that will resonate as true and realistic, enabling teachers to see themselves, the students that they teach, and the parents, caregivers, and communities that they serve as partners on a journey toward mathematics success. Many teachers are already on this path. It is important to understand that the work of achieving this kind of math education is never finished. It is a journey. Different entry points provide opportunities to make progress in various ways but not all at once. The equity-based strategies that you initiate should be grounded in your belief in children's brilliance and connected to your professional goals and teaching vision. We hope that you have gained knowledge to support and enhance your actions toward creating a liberatory and humanizing math education.

5 Equity-Based Practices in Mathematics Classrooms (EquityMath5) Instruction Analysis Tool

PURPOSE: EquityMath5: Instruction Analysis Tool is designed to promote intentional teaching discussions and critical reflection on mathematics lessons with a combined focus on children's mathematical thinking and equity. It is not designed to be an evaluation tool for teachers, but a self-reflective professional tool that can support lesson/unit design and implementation. The EquityMath5 tool aligns with the five equity-based practices discussed in *The Impact of Identity in K–8 Mathematics: Rethinking Equity-Based Practices* (Aguirre, Mayfield-Ingram, and Martin 2013) from the National Council of Teachers of Mathematics.

TOOL DESCRIPTION: The **EquityMath5: Instruction Analysis Tool** consists of five categories of mathematics teaching practices. The categories were adapted from published and unpublished lesson analysis tools and observation protocols (Aguirre, Mayfield-Ingram, and Martin 2013; Aguirre and Zavala 2013; Aguirre, Zavala, and Katanyoutanant 2012; Aguirre et al. 2012; Zavala and Aguirre 2023). Each category connects to a rubric rating scale 1–5 that provides descriptors of instructional practice that affects mathematics identity and experience in the classroom. In addition, there are corresponding reflection prompts to help with lesson analysis. The following table provides a brief description of each category and accompanying reflection prompt.

	Practices	Reflection prompts
1	Going Deep with Mathematics	How does my instruction promote close investigation of math concepts, procedures, and reasoning?
2	Leveraging Multiple Mathematical Competencies	How does my instruction identify and support mathematical contributions from students with varied strengths and levels of confidence?
3	Affirming Mathematics Learning Identities	How does my instruction structure persistence and perseverance with complex math problems?
4	Challenging Spaces of Marginality	How does my instruction widely distribute math knowledge authority in the classroom through centering student-lived experiences as spaces for math investigations, maximizing student-generated math questions, and minimizing status differences among students?
5	Drawing on Multiple Resources of Knowledge (Culture, Language, Family, Community)	How does my instruction affirm and support my students' cultural/family/community-based funds of knowledge and multilingualism to learn mathematics?

HOW TO USE: The best use of this tool is to promote critical discussion and reflection on math lessons with an integrated focus. It is not necessary for every single lesson to have every single category. However, the **EquityMath5: Instruction Analysis Tool** does make explicit the categories of practice that should be consistently evident over time. The tool can be used in the following ways:

1. Analyze videotaped lessons. Individually, rate the lesson enactment based on evidence. Discuss ratings and evidence with a colleague.

2. Examine lesson plans/units. Use evidence to evaluate strengths and areas of growth for equity-based practices.

3. Give general and focused feedback to a peer on an observed lesson.

EquityMath5: Instruction analysis tool

Rating category	1	2	3	4	5
Going deep with mathematics	*How does my instruction promote close analysis of math concepts, procedures, and reasoning?*				
	Instruction is entirely teacher centered, with a sole emphasis on direct instruction.	Instruction is majority teacher centered, with an emphasis on direct instruction.	Instruction includes at least one sustained student-centered activity involving analysis of procedures, concepts, or underlying mathematical structure.	Instruction includes at least half of the lesson with student-centered activities that:	Instruction includes majority student-centered lesson activities that require close analysis of procedures and concepts, involve complex mathematical thinking, utilize multiple representations and strategies, and demand explanation/justification.
	Students receive, recite, or memorize facts, procedures, and definitions without examination.	Students primarily receive, recite, or perform routine procedures without analysis or connection to underlying concepts or mathematical structure.	There is at least one student-centered sustained activity that requires mathematical exploration and analysis.	• Require close analysis of procedures, concepts, or underlying mathematical structure	A large majority of the lesson sustains mathematical analysis.
	One solution strategy is emphasized.	Some opportunities for mathematical exploration and more than one solution strategy, but tasks do not require analysis to complete.	At least two solution strategies are offered.	OR	
	There is no evidence of conceptual understanding being required.			• Require significant mathematical analysis, involve complex mathematical thinking, utilize multiple representations and strategies, or demand explanation/justification	
	There are no opportunities for mathematical analysis or exploration.			There is evidence of sustained mathematical analysis for at least half of the lesson.	

EquityMath5: Instruction analysis tool—continued				
Rating category	1	2	3	4	5
How does my instruction identify and support mathematical contributions from students with varied strengths and levels of confidence?				
Leveraging multiple mathematical competencies	Instruction centers on independent work and individual progress based on predetermined levels of ability.

Student mastery of procedural skills precedes engagement in complex problem solving.

Tasks are closed with virtually no features of mathematical discourse and communication occurring, or what occurs is of a fill-in-the-blank nature with an emphasis on one correct answer or strategy.

Student engagement is low for the majority of the lesson. | Instruction emphasizes individual progress.

Students are grouped by fixed-ability.

Sharing and the development of collective understanding among a few students (or between a single student and the teacher) occur briefly.

Tasks are predominantly closed and highly sequenced with one right solution or strategy emphasized.

Student engagement is predominantly low or sporadic. | Instruction includes one sustained episode of collaborative learning that involves at least one participation structure such as: (1) pair-shares, (2) a small group of mixed-ability students, or (3) a small group of students and the teacher.

OR

Brief episodes of sharing and developing collective understandings occur sporadically throughout the lesson.

At least one task is open with multiple entry points.

Student engagement is mixed. | Instruction includes a majority of activities that foster collaborative learning among students with different strengths.

There are many sustained episodes of sharing and developing collective understandings about mathematics.

The majority of tasks are open with multiple entry points and solution strategies.

Student engagement is high and sustained for the majority of the lesson. | Instruction emphasizes multiple collaborative participation structures that promote sharing of mathematical ideas to solve complex problems.

The creation and maintenance of collective understandings permeate the entire lesson.

All tasks are open with multiple entry points and solution strategies.

Student engagement is high and sustained throughout the lesson. |

EquityMath5: Instruction analysis tool—continued

How does my instruction structure persistence and perseverance with complex math problems?

Rating category	1	2	3	4	5
Affirming mathematics learning identities	Instruction rewards speed, accuracy, and completion over understanding. Explicitly discourages mistakes and immediately corrects them without constructive feedback. Sole focus is on student deficits and errors (what they do not know or cannot do). There is no evidence of valuing flexibility, reasoning, or perseverance.	Instruction emphasizes speed and accuracy with little attention to understanding. Mistakes are tolerated but immediately corrected often without constructive feedback. There is an emphasis on student deficits and errors. There is limited evidence of valuing flexibility, reasoning, and perseverance.	Instruction attends to conceptual understanding, reasoning, and perseverance in at least one part of the lesson. Instruction emphasizes accuracy and efficiency. Mistakes are examined with some attention to partial understandings. Feedback lacks mathematical specificity or only focuses on needs. Instructional attention to both strengths and growth areas is evident in at least one sustained episode of lesson.	Instruction promotes conceptual understanding, persistence, and perseverance in problem solving for sustained periods of time during the lesson. Instruction emphasizes accuracy, flexibility, and efficiency of strategies. Mistakes and errors are sites for learning with opportunities for close analysis. Partial understandings are noted as strengths and opportunities to build understanding. The majority of student feedback is constructive, focusing on math ideas and strategies rather than errors only. Feedback includes strengths and growth areas.	Instruction emphasizes conceptual understanding, perseverance, and reasoning/problem solving throughout the lesson. Accuracy, flexibility, and efficiency are consistently emphasized. Mistakes and errors are sites for learning with opportunities for close analysis. Partial understandings are noted as strengths and opportunities to build understanding. Student feedback is constructive, focuses on math ideas and strategies, addresses strengths and growth areas, and provides opportunities to revise thinking.

EquityMath5: Instruction analysis tool—continued

Rating category	1	2	3	4	5
	How does my instruction widely distribute math knowledge authority in the classroom through centering student-lived experiences as spaces for math investigations, maximizing student-generated math questions, and minimizing status differences among students?				
Challenging spaces of marginality	The authority of math knowledge exclusively resides with the teacher or text. Teacher has final word about correct answers/solutions.	The authority of mathematics knowledge primarily resides with the teacher or text and a few students.	The authority of math knowledge among teachers, text, and students is *sporadically* shared.	The authority of math knowledge is interconnected and shared among teacher, text, and students.	The authority of math knowledge is interconnected and *widely* shared among teacher, text, and students.
	Instruction disconnects or dismisses as inconsequential student-lived experiences and knowledge as sources for math learning.	Instruction attempts to connect to students' lived experiences and knowledge but connections are not meaningful to students.	Instruction makes at least one meaningful connection to students' lived experiences or knowledge.	Instruction facilitates mathematical investigation through multiple meaningful connections to students' lived experiences and knowledge.	Instruction facilitates mathematical investigations through multiple meaningful connections to students' lived experiences and knowledge throughout the lesson and positions students as sources of expertise for solving complex problems.
	Student mathematical contributions are minimal.	Teacher calls on/involves a few students. The mathematical contributions by students are valued and respected.	There is at least one instance where the teacher calls on several students so that multiple mathematical contributions are accepted and valued.	Multiple forms of student mathematical contributions are encouraged and valued.	Student mathematical contributions are actively elicited by the teacher and among students.
	Status differences among students are evident (e.g., segregation of students based on static labels such as (high/low; English learners).	Student involvement is from a particular subgroup (gender, language, ethnicity, class, disability).	The teacher elicits some substantive math contributions.	Teacher and students elicit substantive mathematics contributions.	Multiple strategies to minimize status among students (and specific subgroups) are explicit and widespread throughout the lesson.
	Student voice is restricted to a few, usually high-status (privileged) students.	Status differences among students remain intact and unaddressed.	At least one strategy to minimize status differences among students (and specific subgroups) is evident.	Some strategies to minimize status differences among students (and specific subgroups) throughout the lesson are evident.	

EquityMath5: Instruction analysis tool—continued

Rating category	1	2	3	4	5
Drawing on multiple resources of knowledge (culture, language, family, community)	*How does my instruction affirm and support my students' previous mathematics knowledge, cultural/family/community-based funds of knowledge, and multilingualism to learn mathematics?*				
	Instruction treats previous math knowledge as irrelevant or problematic (e.g., they don't know anything, they lack skills). No evidence of connecting to students' previous mathematical knowledge, thinking, and strategies.	Instruction asserts rather than elicits students' previous math knowledge, thinking, and strategies. Connection to students' previous mathematical knowledge strategies is minimal.	Instruction makes at least one connection to students' previous mathematical knowledge, thinking, or strategies elicited.	Instruction facilitates multiple instances in which students' previous mathematical knowledge, thinking, and strategies are elicited during the lesson.	Level 4-plus instruction extends new knowledge to previous knowledge in multiple ways throughout the lesson.

Mathematics

EquityMath5: Instruction analysis tool—continued

Rating category	1	2	3	4	5
Drawing on multiple resources of knowledge (culture, language, family, community)	*How does my instruction affirm and support my students' previous mathematics knowledge, cultural/family/community-based funds of knowledge, and multilingualism to learn mathematics?*				
Multilingualism	No evidence of a language scaffolding strategy for emerging multilingual students. Students who are new learners of English are ignored and/or seated apart from their classmates. English-only instruction is emphasized. Using other languages is openly discouraged.	Although there is no explicit use of language strategies for emerging multilingual students, students' use of L1 is tolerated. Focus on correct usage and pronunciation of English is emphasized. Discourse is minimized because of perceived English fluency.	There is at least one instance in which a language scaffolding strategy is used to develop academic language (i.e., revoicing; use of cognates; translated tasks/text; use of graphic organizers; strategic grouping with multilingual students). Discourse in multiple languages is encouraged but minimally supported.	Sustained use of at least two language strategies, such as the use of revoicing and attention to cognates, direct modeling of vocabulary, use of realia, strategic grouping of multilingual students, or encouragement of L1 usage, is observed at least between teacher and one student or a small group of students. Discourse in multiple languages is encouraged and supported.	Deliberate and continuous use of language strategies, such as gesturing, use of objects (realia), use of cognates, revoicing, and use of graphic organizers and manipulatives are observed during whole-class and/or small-group instruction and discussions. The main focus is the development of mathematical discourse and meaning making, not students' production of "correct" English.

EquityMath5: Instruction analysis tool—continued

Rating category	1	2	3	4	5
Drawing on multiple resources of knowledge (culture, language, family, community)	*How does my instruction affirm and support my students' previous mathematics knowledge, cultural/family/community-based funds of knowledge, and multilingualism to learn mathematics?*				
Cultural/Community Funds of Knowledge	There is no evidence of connecting to students' cultural funds of knowledge (parental/community knowledge, student interest, histories, lived experiences). OR Lessons build on negative stereotypes of the cultural, community, or family experiences as math resources for lessons (e.g., "Because my kids are Hispanic, I'll use tortillas instead of pizza for fraction lesson").	There is at least one instance in connecting math lessons to students' cultural funds of knowledge (parental/community knowledge, student interest, histories, lived experiences). Lesson draws on student knowledge and experience. Focus is with one student or a small group of students. Lessons incorporate culturally neutral contexts that "all students" will be interested in.	There is at least one sustained episode of sharing and developing collective understanding about mathematics that involves connecting students' cultural funds of knowledge (parental/community knowledge, student interest, histories, lived experiences). OR Brief episodes of sharing and developing collective understandings occur sporadically throughout the lesson.	There are many sustained episodes of sharing and developing collective understandings about mathematics that involve connecting to cultural/community knowledge (e.g., student experiences are mathematized, student-parent connections with math work are encouraged; math examples are embedded in local community/cultural contexts and activities such as games).	The creation and maintenance of collective understandings about mathematics that involve intricate connections to community/cultural knowledge permeate the entire lesson. This would include hook/introduction, main activities, assessment, closure, and homework. Students are asked to analyze the mathematics within the community context and how the mathematics helps them understand and transform that context.

References

Adams, Kimberly S., and Sandra L. Christenson. "Trust and the Family-School Relationship: Examination of Parent-Teacher Differences in Elementary and Secondary Grades." *Journal of School Psychology* 38, no. 5 (2000): 477–97.

Adams Corral, Melissa. "Demanding Equity: Organizing Parents to Fight Tracking." *Medium*, July 30, 2019. https://medium.com/@heinemann/demanding-equity-organizing-parents-to-fight-tracking-6e97e94ce48b.

Adichie, Chimamanda Ngozi, director. "The Danger of a Single Story." TED Conferences LLC , July 2009. www.ted.com/talks/chimamanda_adichie_the_danger_of_a_single_story.

Aguirre, Julia. "Privileging Mathematics and Equity in Teacher Education: Framework, Counter-Resistance Strategies and Reflections from a Latina Mathematics Educator." In *Culturally Responsive Mathematics Education*, edited by Brian Greer, Swapna Mukhopadhyay, Arthur B. Powell, and Sharon Nelson-Barber, pp. 295–319. New York: Routledge, 2009.

———. "Math Strong: Engage, Enrich and Empower Young Mathematical Minds." Invited Presentation. Regional Meeting of the National Council of Teachers of Mathematics. Phoenix, AZ. October 2016.

Aguirre, Julia, Karen Mayfield-Ingram, and Danny Bernard Martin. *The Impact of Identity in K–8 Mathematics Learning and Teaching: Rethinking Equity-Based Practices*. Reston, VA: National Council of Teachers of Mathematics, 2013.

Aguirre, Julia M., Erin E. Turner, Tonya Gau Bartell, Corey Drake, Mary Q. Foote, and Amy Roth McDuffie. "Analyzing Effective Mathematics Lessons for English Learners: A Multiple Mathematical Lens Approach." In *Beyond Good Teaching: Advancing Mathematics Education for ELLs*, edited by Sylvia Celedón-Pattichis and Nora Ramirez, pp. 207–22. Reston, VA: National Council of Teachers of Mathematics, 2012.

Aguirre, Julia M., and Maria del Rosario Zavala. "Making Culturally Responsive Mathematics Teaching Explicit: A Lesson Analysis Tool." *Pedagogies: An International Journal* 8, no. 2 (2013): 163–90.

Aguirre, Julia M., Maria del Rosario Zavala, and Tiffany Katanyoutanant. "Developing Robust Forms of Pre-service Teachers' Pedagogical Content Knowledge through Culturally Responsive Mathematics Teaching Analysis." *Mathematics Teacher Education and Development* 14, no. 2 (2012): 113–36.

American School Counseling Association. "Student to School Counselor Ratio 2021–2022." 2023. https://schoolcounselor.org/About-School-Counseling/School-Counselor-Roles-Ratios. Accessed December 8, 2023.

Anderson, Carol. *Eyes Off the Prize: The United Nations and the African American Struggle for Human Rights, 1944–1955*. Cambridge, UK: Cambridge University Press, 2003.

Anderson, William C., and Zoe Samudzi. *As Black as Resistance: Finding the Conditions for Liberation*. Chico, CA: AK Press, 2018.

Annamma, Subini Ancy, Yolanda Anyon, Nicole M. Joseph, Jordan Farrar, Eldridge Greer, Barbara Downing, and John Simmons. "Black Girls and School Discipline: The Complexities of Being Overrepresented and Understudied." *Urban Education* 54, no. 2 (2016): 211–42.

Annenburg Media. *Valentine Exchange*. Teaching Math: A video library, K–4. No. 42. Burlington, VT: Annenburg Media, 1995.

Apple, Michael W. "Do the Standards Go Far Enough? Power, Policy, and Practice in Mathematics Education." *Journal for Research in Mathematics Education* 23, no. 5 (1992): 412–31.

Archer-Banks, Diane A. M., and Linda S. Behar-Horenstein. "Ogbu Revisited: Unpacking High-Achieving African American Girls' High School Experiences." *Urban Education* 47, no. 1 (2012): 198–223.

Ashlock, Robert B. *Error Patterns in Computation: Using Error Patterns to Improve Instruction*. Upper Saddle River, NJ: Pearson, 2002.

Barajas-López, Filiberto, and Ann M. Ishimaru. "'Darles el lugar': A Place for Nondominant Family Knowing in Educational Equity." *Urban Education* 55, no. 1 (2020): 38–65.

Barajas-López, Filiberto, and Gregory V. Larnell. "Research Commentary: Unpacking the Links between Equitable Teaching Practices and Standards for Mathematical Practice: Equity for Whom and Under What Conditions?" *Journal for Research in Mathematics Education* 50, no. 4 (2019): 349–61.

Bartell, Tonya Gau, Corey Drake, Amy Roth McDuffie, Julia M. Aguirre, Erin E. Turner, and Mary Q. Foote. *Transforming Mathematics Teacher Education*. Cham, Switzerland: Springer International Publishing, 2019.

Beijaard, Douwe, Paulien C. Meijer, and Nico Verloop. "Reconsidering Research on Teachers' Professional Identity." *Teaching and Teacher Education* 20, no. 2 (2004): 107–28.

Bell, Derrick A. Jr. "Brown v. Board of Education and the Interest-Convergence Dilemma." *Harvard Law Review* 93, no. 3 (1980): 518–33.

Berry, Robert Q. III. "Access to Upper-Level Mathematics: The Stories of African American Middle School Boys Who Are Successful with School Mathematics." *Journal for Research in Mathematics Education* 39 (2008): 464–88.

Berry, Robert Q. III, Mark Ellis, and Sherick Hughes. "Examining a History of Failed Reforms and Recent Stories of Success: Mathematics Education and Black Learners of Mathematics in the United States." *Race Ethnicity and Education* 17, no. 4 (2014): 540–68.

Black, Paul, Christine Harrison, Clare Lee, Bethan Marshall, and Dylan Wiliam. "Working inside the Black Box: Assessment for Learning in the Classroom." *Phi Delta Kappan* 86, no. 1 (2004): 9–21.

Blum-Anderson, Judy. "Increasing Enrollment in Higher-Level Mathematics Classes through the Affective Domain." *School Science and Mathematics* 92, no. 8 (1992): 433–36.

Boaler, Jo. *Experiencing School Mathematics: Traditional and Reform Approaches to Teaching and Their Impact on Student Learning*. Rev. ed. Mahwah, NJ: Lawrence Erlbaum, 2002.

———. *Mathematical Mindsets: Unleashing Students' Potential through Creative Math, Inspiring Messages and Innovative Teaching*. Hoboken, NJ: John Wiley & Sons, 2015.

———. *What's Math Got to Do with It? Helping Children Learn to Love Their Least Favorite Subject—and Why It's Important for America*. New York: Viking, 2008.

Boaler, Jo, and Sarah Kate Selling. "Psychological Imprisonment or Intellectual Freedom? A Longitudinal Study of Contrasting School Mathematics Approaches and Their Impact on Adults' Lives." *Journal for Research in Mathematics Education* 48, no. 1 (2017): 78–105.

Borum, Viveka, and Erica Walker. "What Makes the Difference? Black Women's Undergraduate and Graduate Experiences in Mathematics." *The Journal of Negro Education* 81, no. 4 (2012): 366–78.

Brenner, Mary E., and Judit N. Moschkovich. "Everyday and Academic Mathematics in the Classroom." In *Journal for Research in Mathematics Education* Monograph 11. Reston, VA: NCTM, 2002.

Campbell, Shanyce L. "For Colored Girls? Factors That Influence Teacher Recommendations into Advanced Courses for Black Girls." *The Review of Black Political Economy* 39, no. 4 (2012): 389–402.

Christenson, Sandra L. "The Family-School Partnership: An Opportunity to Promote the Learning Competence of All Students." *School Psychology Review* 33, no. 1 (2004): 83–104.

Civil, Marta. "Building on Community Knowledge: An Avenue to Equity in Mathematics Education." In *Improving Access to Mathematics: Diversity and Equity in the Classroom*, edited by Na'ilah Suad Nasir and Paul Cobb, pp. 105–17. New York: Teachers College Press, 2007.

Civil, Marta, and Rosi Andrade. "Collaborative Practice with Parents: The Role of the Researcher as Mediator." In *Collaboration in Teacher Education: Examples from the Context of Mathematics Education*, edited by Andrea Peter-Koop, Vânia Santos-Wagner, C. J. Breen, and A. J. C. Begg, pp. 153–68. Dordrecht, The Netherlands: Kluwer, 2003.

Civil, Marta, and Emily Bernier. "Exploring Images of Parental Participation in Mathematics Education: Challenges and Possibilities." *Mathematical Thinking and Learning* 8, no. 3 (2006): 309–30.

Clements, Douglas H., and Julie Sarama. "Early Childhood Mathematics Learning." In *Second Handbook of Research on Mathematics Teaching and Learning*, edited by Frank K. Lester, pp. 461–555. New York: Information Age, 2007.

Chicago Public Schools. "Fact Sheet on CPS Algebra Exit Exam." https://www.cps.edu/globalassets/cps-pages/about-cps/district-data/metrics/assessment-reports/factsheetoncpsalgebraexitexam.pdf. Accessed April 2023.

Chicago Public Schools. "Algebra Exit Exam Fact Sheet." https://www.cps.edu/academics/student-assessments/ Accessed January 2024.

Coates, Grace Dávila, and Virginia Thompson. *Family Math II: Achieving Success in Mathematics*, K–6. Berkeley, CA: Lawrence Hall of Science, 2003.

Cornell, Stephen E., and Douglas Hartmann. *Ethnicity and Race: Making Identities in a Changing World*. Thousand Oaks, CA: Pine Forge Press, 1998.

Dana Center. *Launch Years: A New Vision for the Transition from High School to Postsecondary Mathematics*. 2020. https://www.utdanacenter.org/sites/default/files/2020-03/Launch-Years-A-New-Vision-report-March-2020.pdf.

Darity Jr., William A., and A. Kirsten Mullen. *From Here to Equality: Reparations for Black Americans in the Twenty-First Century*. Chapel Hill, NC: UNC Press Books, 2022.

Daro, Philip, and Harold Asturias. *Branching Out: Designing Math Pathways for Equity.* Berkeley, CA: Just Equations, 2019. https://uploads-ssl.webflow.com/61 afa2b5ded66610900a0b97/624ccc24db49c225c38c87e4_Just-Equations-2019-Report-Branching-Out-Exec-Summ-Digital.pdf.

Davis, Julius. "A Liberatory Response to Antiblackness and Racism in the Mathematics Education Enterprise." *Canadian Journal of Science, Mathematics and Technology Education* 21, no. 4 (2021): 783–802.

Davis, Julius, and Christopher C. Jett, eds. *Critical Race Theory in Mathematics Education.* New York: Routledge, 2019.

Davis, Julius, and Danny Bernard Martin. "Racism, Assessment, and Instructional Practices: Implications for Mathematics Teachers of African American Students." *Journal of Urban Mathematics Education* 11 (2018): 45–68.

DeCastro-Ambrosetti, Debra, and Grace Cho. "Do Parents Value Education? Teachers' Perceptions of Minority Parents." *Multicultural Education* 13, no. 2 (2005): 44–46.

Desimone, Laura. "Linking Parent Involvement with Student Achievement: Do Race and Income Matter?" *Journal of Educational Research* 93, September/October (1999): 11–30.

Dorner, Lisa M., Marjorie Faulstich Orellana, and Christine P. Li Grining. "'I Helped My Mom, and It Helped Me': Translating the Skills of Language Brokers into Improved Standardized Test Scores." *American Journal of Education* 113, no. 3 (2007): 451–78.

Drake, Corey, James P. Spillane, and Kimberly Hufferd-Ackles. "Storied Identities: Teacher Learning and Subject-Matter Context." *Journal of Curriculum Studies* 33, no. 1 (2011): 1–23.

Dumas, Denis, Daniel McNeish, and Jeffrey A. Greene. "Dynamic Measurement: A Theoretical–Psychometric Paradigm for Modern Educational Psychology." *Educational Psychologist* 55, no. 2 (2020): 88–105.

Ellis, Mark. "Leaving No Child Behind Yet Allowing None Too Far Ahead: Ensuring (in) Equity in Mathematics Education through the Science of Measurement and Instruction." *Teachers College Record* 110, no. 6 (2008): 1330–56.

Epstein, Joyce. "Advances in Family, Community, and School Partnerships." *Community Education Journal* 23, no. 3 (1996): 10–15.

———. "Effects of Teacher Practices of Parent Involvement Change in Student Achievement in Reading and Math." Paper presented at the Annual Meeting of the American Educational Research Association, New Orleans, LA. April 23–27, 1984.

Epstein, R., J. Blake, and T. González. "Girlhood Interrupted: The Erasure of Black Girls' Childhood." Georgetown Law Center on Poverty and Inequality. 2017. https://www.law.georgetown.edu/poverty-inequality-center/wp-content/uploads/sites/14/2017/08/girlhood-interrupted.pdf.

EQUALS and FAMILY MATH. Lawrence Hall of Science, University of California at Berkeley. http://www.lawrencehallofscience.org/equals.

Ernest, Paul. "Empowerment in Mathematics Education." *Philosophy of Mathematics Education Journal* 15, no. 1 (2002): 1–16.

Fabes, Richard A., Matthew Quick, Evandra Catherine, and Aryn Musgrave. 2021. "Exclusionary Discipline in U.S. Public Schools: A Comparative

Examination of Use in Pre-Kindergarten and K-12 Grades." *Educational Studies* ahead-of-print (ahead-of-print). Routledge: 1–18. doi:10.1080/03055698.2021.1941782.

Faulkner, Valerie N., Lee V. Stiff, Patricia L. Marshall, John Nietfeld, and Cathy L. Crossland. "Race and Teacher Evaluations as Predictors of Algebra Placement." *Journal for Research in Mathematics Education* 45, no. 3 (2014): 288–311.

Featherstone, Heather, Sandra Crespo, Lisa M. Jilk, Joy A. Oslund, Amy Noelle Parks, and Marcy B. Wood. *Smarter Together! Collaboration and Equity in the Elementary Math Classroom*. Reston, VA: National Council of Teachers of Mathematics, 2011.

Feldman, Joe. *Grading for Equity: What It Is, Why It Matters, and How It Can Transform Schools and Classrooms*. Thousand Oaks, CA: Corwin Press, 2018.

Flores, Alfinio. "Examining Disparities in Mathematics Education: Achievement Gap or Opportunity Gap?" *The High School Journal* 91, no. 1 (2007): 29–42.

Fuller, Bruce, Edward Bein, Yoonjeon Kim, and Sophia Rabe-Hesketh. "Differing Cognitive Trajectories of Mexican American Toddlers: The Role of Class, Nativity, and Maternal Practices." *Hispanic Journal of Behavioral Sciences* 37, no. 2 (2015): 139–69.

Garcia, A., B. Rincón, and J. Hinojosa. "'There Was Something Missing': How Latinas Construct Compartmentalized Identities in STEM." In *An Asset-Based Approach to Advancing Latina Students in STEM: Increasing Resilience, Participation, and Success*, edited by E. M. Gonzalez, F. Fernandez, and M. Wilson, pp. 181–92. New York: Routledge, 2021.

Gholson, Maisie L. "Clean Corners and Algebra: A Critical Examination of the Constructed Invisibility of Black Girls and Women in Mathematics." *Journal of Negro Education* 85, no. 3 (2016): 290–301.

Gholson, Maisie, and Danny B. Martin. "Smart Girls, Black Girls, Mean Girls, and Bullies: At the Intersection of Identities and the Mediating Role of Young Girls' Social Network in Mathematical Communities of Practice." *Journal of Education* 194, no. 1 (2014): 19–33.

———. "Blackgirl Face: Racialized and Gendered Performativity in Mathematical Contexts." *ZDM* 51 (2019): 391–404.

Gholson, Maisie L., and Charles E. Wilkes. "(Mis) taken Identities: Reclaiming Identities of the 'Collective Black' in Mathematics Education Research through an Exercise in Black Specificity." *Review of Research in Education* 41, no. 1 (2017): 228–52.

Goffney, Imani, and Rochelle Gutiérrez. *Rehumanizing Mathematics for Black, Indigenous, and Latinx Students*. Annual Perspectives in Mathematics Education. Reston, VA: National Council of Teachers of Mathematics, 2018.

González, Norma, Rosi Andrade, Marta Civil, and Luis C. Moll. "Bridging Funds of Distributed Knowledge: Creating Zones of Practices in Mathematics." *Journal of Education for Students Placed at Risk* 6, nos. 1–2 (2001): 115–32.

Gresalfi, Melissa Sommerfeld, and Paul Cobb. "Negotiating Identities for Mathematics Teaching in the Context of Professional Development." *Journal of Mathematics Education* 42, no. 3 (2011): 270–304.

Gresalfi, Melissa S., Taylor Martin, Victoria Hand, and James Greeno. "Constructing Competence: An Analysis of Student Participation in the Activity Systems of Mathematics Classrooms." *Educational Studies in Mathematics* 70, no. 1 (2009): 49–70.

Guskey, Tom. *On Your Mark: Challenging the Conventions of Grading and Reporting.* Bloomington, IN: Solution Tree, 2014.

Gutiérrez, Rochelle. "Beyond Essentialism: The Complexity of Language in Teaching Mathematics to Latina/o Students." *American Educational Research Journal* 39, no. 4 (2002): 1047–88.

———. "Research Commentary: A Gap-Gazing Fetish in Mathematics Education? Problematizing Research on the Achievement Gap." *Journal for Research in Mathematics Education* 39, no. 4 (2008): 357–64.

———. "The Sociopolitical Turn in Mathematics Education." *Journal for Research in Mathematics Education* 44, no. 1 (2013a): 37–68.

———. "Why (Urban) Mathematics Teachers Need Political Knowledge." *Journal of Urban Mathematics Education* 6, no. 2 (2013b): 7–19.

Gutstein, Eric. *Reading and Writing the World with Mathematics: Toward a Pedagogy for Social Justice.* New York: Routledge, 2006.

Gutstein, Eric, and Bob Petersen. *Rethinking Mathematics: Teaching Social Justice by the Numbers.* Milwaukee, WI: Rethinking Schools, 2005.

Heinzman, Erica. "'I Love Math Only if It's Coding': A Case Study of Student Experiences in an Introduction to Data Science Course." *Statistics Education Research Journal* 21, no. 2 (2022): 1–15.

Horn, Ilana. *Strength in Numbers: Collaborative Learning in Secondary Mathematics.* Reston, VA: National Council of Teachers of Mathematics, 2012.

Howard, Nicol. R. "Terms of Engagement: Redefining Parental Involvement and STEM Identity for Black Girls." In *Understanding the Intersections of Race, Gender, and Gifted Education: An Anthology by and about Talented Black Girls and Women in STEM*, edited by N. M. Joseph. Charlotte, NC: Information Age Publishing, 2020.

Howard, Nicol R., Keith E. Howard, R. T. Busse, and Christine Hunt. "Let's Talk: An Examination of Parental Involvement as a Predictor of STEM Achievement in Math for High School Girls." *Urban Education* 58, no. 4 (2023): 586–613.

Howard, Nicol R., and Nicole M. Joseph. "Black Girls in Mathematics:(Re) Envisioning an Inclusive Parent Involvement Measure." *Educational Policy* 36, no. 7 (2022): 1901–28.

Huinker, DeAnn. "Catalyzing Change for Elementary School." *Teaching Children Mathematics* 25, no. 5 (March 2019): 282–88.

Humphreys, Cathy, and Ruth Ann Parker. *Making Number Talks Matter: Developing Mathematical Practices and Deepening Understanding, Grades 3–10.* New York: Routledge, 2015.

Hurtado, Aída, and Patricia Gurin. *Chicana/o Identity in a Changing U.S. Society: Quién Soy? Quiénes Somos?* Tucson, AZ: University of Arizona Press, 2004.

Ishimaru, Ann M. *Just Schools: Building Equitable Collaborations with Families and Communities.* New York: Teachers College Press, 2019.

Ishimaru, Ann M., Filiberto Barajas-López, and Megan Bang. "Centering Family Knowledge to Develop Children's Empowered Mathematics Identities." *Journal of Family Diversity in Education* 1, no. 4 (2015): 1–21.

Jackson, Kara. "The Social Construction of Youth and Mathematics: The Case of a Fifth-Grade Classroom." In *Mathematics Teaching, Learning, and Liberation in the Lives of Black Children*, edited by Danny Bernard Martin, pp. 175–99. New York: Routledge, 2009.

Jackson, K., and J. Remillard. "Rethinking Parent Involvement: African American Mothers Construct Their Roles in the Mathematics Education of Their Children." Scholarly Commons. 2005. https://repository.upenn.edu/gse_pubs/11.

Jones, Stephanie. "Identities of Race, Class, and Gender Inside and Outside the Math Classroom: A Girls' Math Club as a Hybrid Possibility." *Feminist Teacher* (2003): 220–33.

Joseph, Nicole M. *Making Black Girls Count in Math Education: A Black Feminist Vision for Transformative Teaching*. Cambridge, MA: Harvard Education Press. 2022.

Joseph, Nicole M., Meseret Hailu, and Denise Boston. "Black Women's and Girls' Persistence in the P–20 Mathematics Pipeline: Two Decades of Children, Youth, and Adult Education Research." *Review of Research in Education* 41, no. 1 (2017): 203–27.

Joseph, Nicole M., Meseret F. Hailu, and Jamaal Sharif Matthews. "Normalizing Black Girls' Humanity in Mathematics Classrooms." *Harvard Educational Review* 89, no. 1 (2019): 132–55.

King, Natalie S. "Black Girls Matter: A Critical Analysis of Educational Spaces and Call for Community-Based Programs." *Cultural Studies of Science Education* 17, no. 1 (2022): 53–61.

Kirst, M. W., and R. L. Bird. *The Politics of Developing and Maintaining Mathematics and Science Curriculum Content Standards. Research Monograph No. 2.* Madison, WI: University of Wisconsin-Madison, National Institute for Science Education, 1997.

Klein, David. "A Quarter Century of US 'Math Wars' and Political Partisanship." *BSHM Bulletin* 22, no. 1 (2007): 22–33.

Knapp, Andrea, Rachel Landers, and Vetrece Jefferson. *Research Report on Parents and Children in MAAPS*. Athens, GA: University of Georgia, 2012.

Lampert, Magdalene. *Teaching Problems and the Problems of Teaching*. New Haven, CT: Yale University Press, 2001.

Langer-Osuna, Jennifer Marie, and Indigo Esmonde. "Identity in Research on Mathematics Education." In *Compendium for Research in Mathematics Education*, edited by J. Cai, pp. 637–48. Reston, VA: National Council of Teachers of Mathematics, 2017.

Larnell, Gregory V. "More Than Just Skill: Examining Mathematics Identities, Racialized Narratives, and Remediation among Black Undergraduates." *Journal for Research in Mathematics Education* 47, no. 3 (2016): 233–69.

Lee, Stacey J. *Up Against Whiteness: Race, School, and Immigrant Youth*. New York: Teachers College Press, 2005.

———. *Unraveling the "Model Minority" Stereotype: Listening to Asian American Youth.* 2nd ed. New York: Teachers College Press, 2009.

Leonard, Jacqueline. "Black Lives Matter in Teaching Mathematics for Social Justice." *Journal of Urban Mathematics Education* 13 (2020): 5–11.

Leonard, Jacqueline, and Danny B. Martin, eds. *The Brilliance of Black Children in Mathematics*. Charlotte, NC: IAP, 2013.

Leong, Nancy. "Racial Capitalism." *Harvard Law Review* 126 (2013): 2151–226.

Lester, Frank K., ed. *Second Handbook of Research on Mathematics Teaching and Learning: A Project of the National Council of Teachers of Mathematics*. Charlotte, NC: IAP, 2007.

Lewis, Amanda E., John B. Diamond, and Tyrone A. Forman. "Conundrums of Integration: Desegregation in the Context of Racialized Hierarchy." *Sociology of Race and Ethnicity* 1, no. 1 (2015): 22–36.

Louie, V. (2004). "Compelled to Excel: Immigration, Education, and Opportunity among Chinese Americans." Redwood City, CA: Stanford University Press.

MacArthur, K. "Re-Humanizing Assessments in University Calculus II Courses." In *Proceedings of the 22nd Annual Conference on Research in Undergraduate Mathematics Education*, edited by Aaron Weinberg, Deborah Moore-Russo, Hortensia Soto, and Megan Wawro, pp. 953–8. 2019.

Lyken-Segosebe, Dawn, and Serena E. Hinz. "The Politics of Parental Involvement: How Opportunity Hoarding and Prying Shape Educational Opportunity." *Peabody Journal of Education* 90, no. 1 (2015): 93–112.

Manyak, Patrick C. "'What Did She Say?': Translation in a Primary-Grade English Immersion Class." *Multicultural Perspectives* 6, no. 1 (2004): 12–18.

Martin, Danny B. "Equity, Inclusion, and Antiblackness in Mathematics Education." *Race Ethnicity and Education* 22, no. 4 (2019): 459–78.

———. "Liberating the Production of Knowledge about African American Children and Mathematics." In *Mathematics Teaching, Learning, and Liberation in the Lives of Black Children*, edited by Danny Bernard Martin, pp. 3–36. New York: Routledge, 2009a.

———. "Mathematics Learning and Participation as Racialized Forms of Experiences: African-American Parents Speak on the Struggle for Mathematics Literacy." *Mathematical Thinking & Learning* 8, no. 3 (2006): 197–229.

———. *Mathematics Success and Failure among African American Youth: The Roles of Sociohistorical Context, Community Forces, School Influence, and Individual Agency*. Mahwah, NJ: Lawrence Erlbaum, 2000.

———. "Researching Race in Mathematics Education." *Teachers College Record* 111, no. 2 (2009b): 295–338.

Martin, Danny Bernard, Paula Groves Price, and Roxanne Moore. "Refusing Systemic Violence against Black Children: Toward a Black Liberatory Mathematics Education." In *Critical Race Theory in Mathematics Education*, edited by Julius Davis and Christopher C. Jett, pp. 32–55. New York: Routledge, 2019.

Math and Parent Partners (MAPPS). University of Arizona. https://mappsua .wordpress.com.

Mathematics Assessment Resource Service (MARS). Silicon Valley Mathematics Initiative, 2010. https://www.mathshell.org/ba_mars.htm.

Mayfield-Ingram, Karen, and Alma Ramirez. *The Journey—Through Middle School Math.* Berkeley, CA: EQUALS Programs, Lawrence Hall of Science, University of California at Berkeley, 2005.

McGee, Ebony, and Margaret Beale Spencer. "Black Parents as Advocates, Motivators, and Teachers of Mathematics." *Journal of Negro Education* 84, no. 3 (2015): 473–90.

McGee, E. O., and F. A. Pearman. "Understanding Black Male Mathematics High Achievers from the Inside Out: Internal Risk and Protective Factors in High School." *Urban Review* 47, no. 3 (2015): 513–40. doi:10.1007/s11256-014-0317-2.

Migration Policy Institute. "Children in U.S. Immigrant Families, 2018." https://www.migrationpolicy.org/programs/data-hub/charts/children-immigrant-families. Accessed June 24, 2020.

Miller-Cotto, D., and N. A. Lewis Jr. "Am I a 'Math Person'? How Classroom Cultures Shape Math Identity among Black and Latinx Students, 2020." OSF Preprints. https://osf.io/hcqst.

Miller-Jones, Dalton, and Brian Greer. "Conceptions of Assessment of Mathematical Proficiency and Their Implications for Cultural Diversity." In *Culturally Responsive Mathematics Education*, edited by Brian Greer, Swapna Mukhopadhyay, Arthur B. Powell, and Sharon Nelson-Barber, 165–86. New York: Routledge, 2009.

Milner, H. Richard IV. "Analyzing Poverty, Learning, and Teaching through a Critical Race Theory Lens." *Review of Research in Education* 37, no. 1 (2013): 1–53.

Minke, Kathleen M., and Kellie J. Anderson. "Restructuring Routine Parent-Teacher Conferences: The Family-School Conference Model." *Elementary School Journal* 104, no. 1 (2003): 49–69.

Morris, Edward W. "'Ladies' or 'Loudies'? Perceptions and Experiences of Black Girls in Classrooms." *Youth & Society* 38, no. 4 (2007): 490–515.

Morris, Monique. *Pushout: The Criminalization of Black Girls in Schools.* New York: The New Press, 2016.

Morton, C. H., and D. Smith-Mutegi. "Girls STEM Institute: Transforming and Empowering Black Girls in Mathematics through STEM." In *Rehumanizing Mathematics for Black, Indigenous, and Latinx Students*, edited by I. Goffney and R. Gutiérrez, pp. 233–37. Reston, VA: National Council of Teachers of Mathematics, 2018.

Morton, C., D. Tate McMillan, and W. Harrison-Jones. "Black Girls and Mathematics Learning." Oxford Research Encyclopedia of Education. July 30, 2020. https://oxfordre.com/education/view/10.1093/acrefore/9780190264093.001.0001/acrefore-9780190264093-e-1028. Accessed December 7, 2023.

Moschkovich, Judit N. "Learning Mathematics in Two Languages: Moving from Obstacles to Resources." In *Changing the Faces of Mathematics: Perspectives on Multiculturalism and Gender Equity*, edited by Walter Secada, pp. 85–93. Reston, VA: National Council of Teachers of Mathematics, 1999.

———. "A Situated and Sociocultural Perspective on Bilingual Mathematics Learners." *Mathematical Thinking and Learning* 4, nos. 2–3 (2002): 189–212.

Nasir, Na'ilah. *Racialized Identities: Race and Achievement among African American Youth.* Redwood City, CA: Stanford University Press, 2011.

National Center for Educational Statistics. "Digest of Educational Statistics." 2021. https://nces.ed.gov/programs/digest/d21/tables/dt21_204.10.asp. Accessed April 25, 2023.

National Center for Education Statistics. "Racial/Ethnic Enrollment in Public Schools." Condition of Education. U.S. Department of Education, Institute of Education Sciences. 2022. https://nces.ed.gov/programs/coe/indicator/cge. Accessed April 25, 2023.

National Center for Science and Engineering Statistics. Data Tables. 2020. https://ncsesdata.nsf.gov/sere/2018/html/sere18-dt-tab006.html. Accessed June 24, 2020.

National Council of Teachers of Mathematics (NCTM). *Assessment Standards for School Mathematics.* Reston, VA: NCTM, 1995.

———. *Catalyzing Change in High School Mathematics: Initiating Critical Conversations.* Reston, VA: Author, 2018.

———. *Curriculum and Evaluation Standards for School Mathematics.* Reston, VA: Author, 1989.

———. *Principles to Actions: Ensuring Mathematics Success for All.* Reston, VA: Author, 2014.

———. *Principles and Standards for School Mathematics.* Reston, VA: Author, 2000.

National Governors Association Center for Best Practices and Council of Chief State School Officers (NGA Center and CCSSO). "Common Core State Standards." Washington, DC: NGA Center and CCSSO, 2010. http://www.corestandards.org.

National Research Council. *Adding It Up: Helping Children Learn Mathematics*, edited by Mathematics Learning Study Committee, Jeremy Kilpatrick, Jane Swafford, and Bradford Findell. Center for Education, Division of Behavioral and Social Sciences and Education. Washington, DC: National Academy Press, 2001a.

———. *Improving Mathematics Education: Resources for Decision Making*, edited by Committee on Decisions That Count, Steve Leinwand and Gail Burrill. Washington, DC: National Academy Press, 2001b.

Oakes, Jeannie. *Keeping Track: How Schools Structure Inequality.* 2nd ed. New Haven, CT: Yale University Press, 2005.

Olsen, Brad. "'I Am Large, I Contain Multitudes': Teacher Identity as a Useful Frame for Research, Practice, and Diversity in Teacher Education." In *Studying Diversity in Teacher Education*, edited by Arnetha F. Ball and Cynthia A. Tyson, pp. 257–73. Lanham, MD.: Rowman & Littlefield, 2011.

Orellana, Marjorie Faulstich. *Translating Childhoods: Immigrant Youth, Language, and Culture.* Piscataway, NJ: Rutgers University Press, 2009.

Orfield, G. "The Oak Park Way Isn't Enough." Lasting Diversity Achieved. School Challenges Remaining in Suburban Chicago. In *The Resegregation of Suburban Schools: A Hidden Crisis in American Education*, edited by E. Frankenberg and G. Orfield, pp. 185–214. Cambridge, MA: Harvard Education Press, 2012.

Parrish, Sherry D. *Number Talks: Helping Children Build Mental Math and Computation Strategies, Grades K–5.* Sausalito, CA: Math Solutions, 2010.

Peressini, Dominic D. "The Portrayal of Parents in the School Mathematics Reform Literature: Locating the Context for Parental Involvement." *Journal for Research in Mathematics Education* 29 (1998): 555–82.

Remillard, Janine T., and Kara Jackson. "Old Math, New Math: Parents' Experiences with Standards-Based Reform." *Mathematical Thinking & Learning* 8, no. 3 (2006): 231–59.

Robinson, C. (2000 [1983]). *Black Marxism: The making of the black radical tradition.* Chapel Hill: University of North Carolina Press.

Romberg, Thomas A. "Further Thoughts on the Standards: A Reaction to Apple." *Journal for Research in Mathematics Education* 23, no. 5 (1992): 432–37.

Russell, Susan Jo. "Developing Computational Fluency with Whole Numbers." *Teaching Children Mathematics* 7, no. 3 (2000): 154–58.

Sanchez, Claudio. "Mexican-American Toddlers: Understanding The Achievement Gap." National Public Radio. April 7, 2015. https://www.npr.org/sections/ed/2015/04/07/397829916/mexican-american-toddlers-understanding-the-achievement-gap. Accessed April 2015.

Schoenfeld, Alan H. "Learning to Think Mathematically: Problem Solving, Metacognition, and Sense Making in Mathematics." In *Handbook of Research on Mathematics Teaching and Learning*, edited by Douglas A. Grouws, pp. 334–70. New York: Macmillan; Reston, VA: National Council of Teachers of Mathematics, 1992.

———. "The Math Wars." *Educational Policy* 18, no. 1 (2004): 253–86.

Shah, Niral. "'Asians Are Good at Math' Is Not a Compliment: STEM Success as a Threat to Personhood." *Harvard Educational Review* 89, no. 4 (2019): 661–86.

Simic-Muller, Ksenjia, Erin E. Turner, and Maura C. Varley. "Math Club Problem-Posing." *Teaching Children Mathematics* 16, no. 4 (2009): 206–12.

Smith Arrillaga, E., S. Bland, K. Goto, and M. Almora Rios. "Integral Voices: Examining Math Experiences of Underrepresented Students." *Just Equations.* 2023. https://justequations.org/resource/integral-voices-examining-math-experiences-of-underrepresented-students.

Spielhagen, Frances R. *The Algebra Solution to Mathematics Reform: Completing the Equation.* New York: Teachers College Press, 2011.

Staats, Susan. "Somali Mathematics Terminology: A Community Exploration of Mathematics and Culture." In *Multilingualism in Mathematics Classrooms: Global Perspectives*, edited by Richard Barwell, pp. 32–46. Bristol, CT: Multilingual Matters, 2009.

Stein, Mary Kay, Margaret Schwan Smith, Marjorie A. Henningsen, and Edward A. Silver. *Implementing Standards-Based Mathematics Instruction: A Casebook for Professional Development.* New York: Teachers College Press, 2000.

Stiggens, Richard. *The Perfect Assessment System.* Alexandria, VA: Association for Supervision and Curriculum Development, 2017.

Stinson, David. "Negotiating Sociocultural Discourses: The Counter-Storytelling of Academically (and Mathematically) Successful African American Male Students." *American Educational Research Journal* 45, no. 4 (2008): 975–1010.

Stipek, Deborah, Karen B. Givvin, Julie M. Salmon, and Valanne L. MacGyvers. "Teacher Beliefs and Practices Related to Mathematics Instruction." *Teaching and Teacher Education* 17 (February 2001): 213–26.

Tate, William F. "Mathematics Standards and Urban Education: Is This the Road to Recovery?" *The Educational Forum* 58, no. 4 (1994a): 380–390.

———. "Race, Retrenchment, and Reform of School Mathematics." *Phi Delta Kappan* 75, no. 6 (1994b): 477–84.

———. "School Mathematics and African American Students: Thinking Seriously about Opportunity-to-Learn Standards." *Educational Administration Quarterly* 31, no. 3 (1995): 424–48.

———. "What Is a Standard?" *Proceedings of the NCTM Research Catalyst Conference*, edited by F. K. Lester and J. Ferrini-Mundy, pp. 15–23. Reston, VA: National Council of Teachers of Mathematics, 2004.

Terman, Lewis Madison. *The Measurement of Intelligence: An Explanation of and a Complete Guide for the Use of the Stanford Revision and Extension of the Binet-Simon Intelligence Scale.* Boston, MA: Houghton Mifflin, 1916.

Texas Essential Knowledge and Skills for Kindergarten—Grade 12: 19 TAC. Chapter 111, Mathematics. Austin, TX: Texas Education Agency, 2012.

Torres, Angela, Ho Nguyen, Elizabeth Hull Barnes, and Laura Wentworth. *A Guide to Detracking Math Courses: The Journey to Realize Equity and Access in K-12 Mathematics Education.* Corwin Press, 2023.

Turner, Erin. "Critical Mathematical Agency: Urban Middle School Students Engage in Significant Mathematics to Understand, Critique, and Act upon Their World." PhD diss., University of Texas, 2003.

Turner, Erin, and Sylvia Celedón-Pattichis. "Problem Solving and Mathematical Discourse among Latino/a Kindergarten Students: An Analysis of Opportunities to Learn." *Journal of Latinos and Education* 10, no. 2 (2011): 146–69.

Turner, Erin E., Corey Drake, Amy Roth McDuffie, Julia M. Aguirre, Tonya Gau Bartell, and Mary Q. Foote. "Promoting Equity in Mathematics Teacher Preparation: A Framework for Advancing Teacher Learning of Children's Multiple Mathematics Knowledge Bases." *Journal of Mathematics Teacher Education* 15, no. 1 (2012): 67–82.

Turner, Erin E., and Beatriz T. Font Strawhun. "Posing Problems That Matter: Investigating School Overcrowding." *Teaching Children Mathematics* 13, no. 9 (2007): 457–63.

University of Chicago School Mathematics Project. *Everyday Mathematics.* Chicago: McGraw Hill, 2007.

Valencia, Richard R. *Dismantling Contemporary Deficit Thinking: Educational Thought and Practice.* New York: Routledge, 2010.

Washington, Delaina G. 2019. "Pour Out the Oil: Successful Parenting for Math Development among African Americans." Order No. 27987797, University of Illinois at Chicago. https://www.proquest.com/dissertations-theses/pour-out-oil-successful-parenting-math/docview/2405150465/se-2.

Washington, D., Z. Torres, M. Gholson, and D. B. Martin. "Crisis as a Discursive Frame in Mathematics Education Research and Reform: Implications for Educating Black Children." In *Alternative Forms of Knowing (in) Mathematics*, edited by S. Mukhopadhay and W.M. Roth, pp. 53–70. Rotterdam, The Netherlands: Sense Publishers, 2012.

Watanabe, Teresa. "Literacy Gap between Latino and White Toddlers Starts Early, Study Shows" *Los Angeles Times*. https://www.latimes.com/local/lanow/la-me-ln-latino-literacy-20150401-story.html#:~:text=The%20UC%20Berkeley%20study%20found,and%20familiarity%20with%20print%20materials. Accessed April 2015.

Wells, Cacey L. "Understanding Issues Associated with Tracking Students in Mathematics Education." *Journal of Mathematics Education* 11, no. 2 (2018): 68–84.

Wilkes II, Charles. "Ordinary Brilliance: Understanding Black Children's Conceptions of Smartness and How Teachers Communicate Smartness through Their Practice." Diss., University of Michigan, 2022. https://deepblue.lib.umich.edu/handle/2027.42/172637.

Wood, Marcy B. "Mathematical Micro-Identities: Moment-to-Moment Positioning and Learning in a Fourth-Grade Classroom." *Journal for Research in Mathematics Education* 44, no. 5 (2013): 775–808.

Yan, Wenfan, and Qiuyun Lin. "Parent Involvement and Mathematics Achievement: Contrast across Racial and Ethnic Groups." *Journal of Educational Research* 99, no. 2 (2005): 116.

Yeager, David S., Valerie Purdie-Vaughns, Julio Garcia, Nancy Apfel, Pattie Brzustoski, Allison Master, William T. Hessert, Matthew E. Williams, and Geoffrey L. Cohen. "Breaking the Cycle of Mistrust: Wise Interventions to Provide Critical Feedback across the Racial Divide." *Journal of Experimental Psychology: General* 143, no. 2 (2014): 804–24. https://doi.org/10.1037/a0033906.

Yeh, Cathery, and Brande M. Otis. "Mathematics for Whom: Reframing and Humanizing Mathematics." *Occasional Paper Series* 2019, no. 41 (2019): 8.

Zavala, Maria del Rosario, and Julia M. Aguirre. *Cultivating Mathematical Hearts: Culturally Responsive Mathematics Teaching in Elementary Classrooms.* Thousand Oaks, CA: Corwin Press, 2023.